DEDUCTIVE GEOMETRY

E. A. MAXWELL

DOVER PUBLICATIONS, INC.
Mineola, New York

Bibliographical Note

This Dover edition, first published in 2015, is an unabridged republication of the work originally published in 1962 by the Pergamon Press, Oxford and New York.

Library of Congress Cataloging-in-Publication Data

Maxwell, E. A. (Edwin Arthur)
 Deductive geometry / E. A. Maxwell.—Dover edition.
    p. cm.
 Originally published: Oxford: Pergamon Press, 1962.
 Includes index.
 ISBN-13: 978-0-486-80283-1
 ISBN-10: 0-486-80283-3
 1. Geometry. I. Title.

QA447.M39 2015
516.2—dc23

2015028581

Manufactured in the United States by RR Donnelley
80283301   2015
www.doverpublications.com

Contents

PREFACE vii

Chapter 1 · INTRODUCTION AND NOTATION 1

Chapter 2 · THE GEOMETRY OF THE TRIANGLE 14

Chapter 3 · SOME CIRCLE THEOREMS 27

Chapter 4 · THE THEOREMS OF CEVA AND MENELAUS 51

Chapter 5 · HARMONIC PROPERTIES 57

Chapter 6 · POLE AND POLAR 78

Chapter 7 · LINE AND PLANE 90

Chapter 8 · SOME STANDARD SOLID BODIES 107

Chapter 9 · ANGLES BETWEEN LINES AND PLANES 120

Chapter 10 · THE SPHERE 130

Chapter 11 · THE NATURE OF SPACE 144

Chapter 12 · TRANSFORMATIONS 155

INDEX 175

Preface

THE AIM of this book is to give, in concise form, the whole of the geometry of the straight line, circle, plane and sphere, with their associated configurations such as triangle or cylinder, in so far as it is likely to be required for courses in mathematics in the United Kingdom for the G.C.E. at Advanced and Scholarship levels, or for corresponding courses throughout the Commonwealth as required by the appropriate examining boards. The book will be of value also to university undergraduates.

This is a subject which, at the moment of writing, is less popular than it deserves, but I hope that the treatment may help to stimulate interest as well as to satisfy an existing need.

The plan of the book is straightforward; recapitulation of known work, advanced plane geometry, solid geometry with some reference to the geometry of the sphere, a chapter on the nature of space with reference to such properties as congruence, similarity and symmetry, and, finally, a very brief account of the elementary transformations of projection and inversion.

The book is interspersed with a number of examples. Bearing in mind the need for brevity, a number of these are actually standard results which the reader is invited to prove for himself. Such examples are headed *Theorems*.

I would express my thanks to Dr. H. M. Cundy for many valuable suggestions, and also to the staff of Pergamon Press for all their trouble and skill.

E. A. Maxwell
Fellow of Queens College, Cambridge

ONE
Introduction and Notation

THE READER who comes to this book is expected to be familiar with the normal concepts of elementary geometry as commonly taught at school: length and angle; similarity and congruence; point, line and circle; area and the theorem of Pythagoras. Such knowledge will, presumably, rest on an empirical basis, leading to an appreciation of the standard theorems and of the general structure of geometrical argument, but without that detailed investigation which was prevalent until the start of this century or even later.

It has recently been realized that the present lack of training in geometrical argument must, for the young student of mathematics, be corrected in some way unless his ability to handle formal mathematical work is to be endangered. A strong candidate for the purpose is formal algebra, which is to be welcomed wholeheartedly. This book seeks to achieve a similar end, but using as alternative subject-matter some topics in geometry which are usually studied in the upper school.

The book will, however, have a somewhat strange look, even to those who are completely familiar with the material, for it is presented in the notation that is in regular use in more modern work in mathematics. *It is emphasized that the number of new symbols is small and that they are introduced not only to serve the purposes of this book but also to help pupils to become familiar with their use elsewhere.* Experience seems to indicate that, at about the upper school stage, pupils develop a positive enthusiasm for new symbolism, especially when it helps to reduce the burden of writing, and it is hoped that they will readily respond to this approach.

One other procedural innovation should be mentioned. By long tradition, geometrical arguments have been set out under the formal headings, *Given, Required, Construction, Proof*. The discipline has much to commend it, but it is harder to sustain as work progresses; on the other hand, a recognizable structure is helpful both to writer and to reader. In so far as it is possible, therefore, the treatment of each property will begin with a statement: *The Problem*, and this will be followed by a proof under the heading: *The Discussion*.

1. Standard Notation

The following standard notation of elementary geometry will be used regularly:

$\angle ABC$ or $\angle B$	the angle ABC
$\triangle ABC$	the triangle ABC
$\triangle ABC \equiv \triangle PQR$	the triangles are congruent, *with the implication* $BC = QR$, $CA = RP$, $AB = PQ$
$\triangle ABC \sim \triangle PQR$	the triangles are similar, *with the implication* $\angle A = \angle P$, $\angle B = \angle Q$, $\angle C = \angle R$
$AB = PQ$	the lengths AB, PQ are equal
$AB \perp PQ$	AB is perpendicular to PQ†

† When the context makes it clear, we assume without explicit statement that "$AB \perp PQ$" means that B is the foot of the perpendicular from A to PQ.

$AB \| PQ$ AB is parallel to PQ
$\odot ABC$ the circle ABC

2. Fresh Notation

The notation explained in this paragraph, though now in common usage, will almost certainly be new to pupils in upper schools.

(i) THE SYMBOL OF INCLUSION $\in$. The symbol $\in$ is used in the sense that "$P \in l$" means, "P is included among those elements which constitute the set of elements l".

In a geometrical context, the statement might mean, "P is a point of the line l".

In practice, a line is often named in terms of two of its points A and B. We then write "$P \in AB$".

(ii) THE SYMBOL OF UNION $\cup$. The symbol $\cup$ is used in the sense that "$AB \cup CD$" means, "all the points which belong to AB, to CD, or to both". It thus *unites* into the single entity $AB \cup CD$ those points which belong to AB or CD severally.

(This symbol will, in fact, not be used in this book, but it is introduced here because the symbol "$\cup$" for "union" is natural whereas the next symbol, used more often, is less self-explanatory. Care must be taken not to confuse the two, and the mnemonic "$\cup$ for union" is helpful for this.)

(iii) THE SYMBOL OF INTERSECTION $\cap$. The symbol $\cap$ is used in the sense that "$AB \cap CD$" means, "all the points which belong both to AB and to CD".

For example, it is an immediate consequence of the definitions that

$$AB \cap CD \in AB$$

That is, the intersection of AB and CD belongs to AB.

(The symbols "$AB \cup CD$" and "$AB \cap CD$" are sometimes, for obvious reasons, read as "AB cup CD" and "AB cap CD".)

(iv) THE SYMBOL OF CONSEQUENCE $\Rightarrow$. The symbol $\Rightarrow$ is used in the sense that a chain of argument like

$$"2x + 5 = 7x + 14$$
$$\Rightarrow \quad 5x = -9$$
$$\Rightarrow \quad x = -\tfrac{9}{5}"$$

means,

"the equation $2x + 5 = 7x + 4$
leads to $\quad 5x = -9$
and this *leads to* $\quad x = -\tfrac{9}{5}$"

The important thing about this symbol is the way the arrow points. The first statement *leads to* the second. Care should be taken not to reverse the argument without ensuring that such reversal is legitimate. For example:

$ABCD$ is a rectangle
$\Rightarrow AB = DC$ and $AD = BC$;

but it is *not true* that

$AB = DC$ and $AD = BC$
$\Rightarrow ABCD$ is a rectangle.

On the other hand, both of the statements

$$AB = AC \Rightarrow \angle ACB = \angle ABC$$

and

$$\angle ACB = \angle ABC \Rightarrow AB = AC$$

are true. The notation

$$AB = AC \Leftrightarrow \angle ACB = \angle ABC$$

is often used to denote this fact.

It may be useful to digress for a moment to emphasize one or two points which are probably familiar. Consider, for example, the theorem:
"In $\triangle ABC$, $\triangle PQR$,
 $BC = QR$, $CA = RP$, $AB = PQ$
 $\Rightarrow \angle BAC = \angle QPR, \angle CBA = \angle RQP, \angle ACB = \angle PRQ$."

The facts *sides equal* and *angles equal* do indeed "go together" in a sense, but the argument

sides equal $\Rightarrow$ angles equal

cannot be reversed: it is NOT true that

angles equal $\Rightarrow$ sides equal.

In other words, it is *not legitimate* to interchange the rôles of data and conclusion in an argument without careful examination.

Definition. A result obtained from a given theorem by interchanging rôles of data and conclusion is called a *converse* of that theorem.

The point we have been making is that *a converse of a true theorem is not necessarily true*.

The symbol $\Leftrightarrow$ to which we have just referred may be used only when theorem and converse are both true. For example:

ABCD is a cyclic quadrilateral
$\Leftrightarrow$ the sum of opposite angles is $180°$.

Analogous ideas in common use depend on the use of the word "*if*". The reader is strongly urged to be careful to use this very simple word completely unambiguously. A convenient way to be quite sure is to use the verbal formula "*if* . . . *then* . . .". For example:

If ABCD is a cyclic quadrilateral, *then* the sum of opposite angles is $180°$.

In this example, the converse is also true:

If the sum of the opposite angles of the quadrilateral *ABCD* is $180°$, *then* the quadrilateral is cyclic.

Note how the formula "*if* . . . *then* . . ." follows the sense of the arrow ⇒.

When both theorem and converse are true, the phrase "*if and only if*" is often used:

The quadrilateral *ABCD* is cyclic *if and only if* the sum of the opposite angles is 180°.

Thus the formula "*if and only if*" follows the two senses of the double arrow ⇔. The statement so enunciated implies two distinct problems, which often need separate solution.

In many of the arguments which follow, there are steps where the double symbol ⇔ might be applied legitimately but where only the sense ⇒ is relevant. In such cases, the single symbol ⇒ is usually adopted.

When two or more conditions must be taken together to lead to a result, they will often be linked by a bracket. Thus:

$$\left.\begin{array}{l} AP = PQ, AC = PR \\ \angle BAC = \angle QPR \end{array}\right\} \Rightarrow \triangle ABC \equiv \triangle PQR \Rightarrow BC = QR.$$

(v) THE SYMBOL OF EXISTENCE ∃. The symbol ∃ is used in the sense that

"∃ O such that $OA = OB = OC$"

means, "there exists a point O—the circumcentre of the triangle ABC—such that $OA = OB = OC$".

Remark. The notation just introduced is, at this stage, merely a notation, with which the reader is expected to become familiar quickly. It use carries no implications of set theory or symbolic logic, though knowledge of it may help when these subjects come forward for study later.

3. Results Assumed Known

The following summary indicates the results on which arguments will be founded and also gives an introduction to some of the notation just explained. For brevity, plentiful use is made of diagrams.

Introduction and Notation

(i) ANGLE PROPERTIES AND PARALLEL LINES.

AUB straight line $\Leftrightarrow \angle AUP + \angle PUB = 180°$.
AUB straight line $\Leftrightarrow \angle AUP = \angle BUV$.
$CD \| AB \Leftrightarrow \angle AUP = \angle CVP$,
$CD \| AB \Leftrightarrow \angle AUV = \angle UVD$,
$CD \| AB \Leftrightarrow \angle AUV + \angle UVC = 180°$.

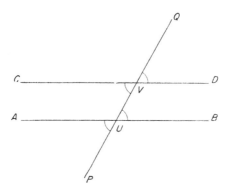

Fig. 1

(ii) ANGLE PROPERTIES FOR A TRIANGLE.

$\angle A + \angle B + \angle C = 180°$,
$\angle ACD = \angle A + \angle B$.

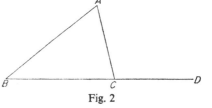

Fig. 2

Extension. The sum of the angles of a polygon of n sides is $2n - 4$ right angles; in particular, the sum of the angles of a quadrilateral is $360°$.

(iii) CONGRUENCE OF TRIANGLES.

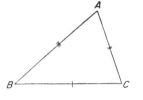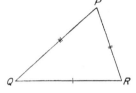

Fig. 3

$\triangle ABC \equiv \triangle PQR$
$\Leftrightarrow AB = PQ, AC = PR, \angle A = \angle P,$
$\Leftrightarrow BC = QR, CA = RP, AB = PQ,$
$\Leftrightarrow BC = QR, \angle B = \angle Q, \angle C = \angle R.$

Note the double sense $\Leftrightarrow$ of the arrows.

Note, too, that $\angle A = \angle P, \angle B = \angle Q, \angle C = \angle R \not\Rightarrow \triangle ABC \equiv \triangle PQR$; three equal angles are *not enough* for congruence.

(iv) SIMILARITY OF TRIANGLES.

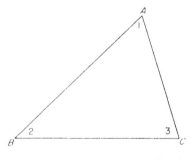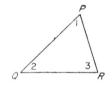

Fig. 4

$\triangle ABC \sim \triangle PQR$
$\Leftrightarrow \angle A = \angle P, \angle B = \angle Q, \angle C = \angle R$

$$\Leftrightarrow \frac{BC}{QR} = \frac{CA}{RP} = \frac{AB}{PQ},$$
$$\Leftrightarrow \angle A = \angle P, \ \frac{BA}{QP} = \frac{CA}{RP}.$$

Note. As a particular case,

$UV \| BC$
$\Leftrightarrow \dfrac{AU}{AB} = \dfrac{AV}{AC}$
$\Leftrightarrow \dfrac{AU}{UB} = \dfrac{AV}{VC}$

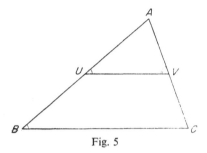

Fig. 5

Still more particularly,

$\left. \begin{array}{l} U \text{ middle point of } AB \\ V \text{ middle point of } BC \end{array} \right\} \Rightarrow UV \| BC \text{ and } UV = \tfrac{1}{2} BC.$

(v) AREA.

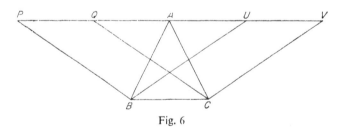

Fig. 6

$PQAUV \| BC$
$\Leftrightarrow$ area $BCQP =$ area $BCVU$
$\qquad\qquad = 2 \triangle ABC.$

(vi) THE THEOREM OF PYTHAGORAS.

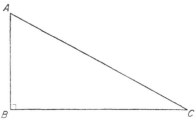

Fig. 7

$$\angle ABC = 90°$$
$$\Leftrightarrow AC^2 = AB^2 + BC^2.$$

(vii) THE TRIANGLE INEQUALITY.

The sum of two sides of a triangle is greater than the third side. In symbols,

$$AB + AC > BC, \quad BC + BA > CA, \quad CA + CB > AB.$$

Note. $AB + AC = BC \Rightarrow A, B, C$ are collinear.

(viii) SYMMETRY PROPERTY OF A CIRCLE.
 $AU = UB$
 $\Rightarrow OU \perp AB.$

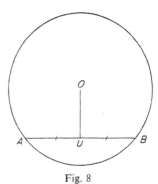

Fig. 8

(ix) Angle Properties of a Circle.

$\angle AOB = 2\angle AUB$.

$ABVU$ cyclic
$\Leftrightarrow \angle AUB = \angle AVB$.

$AWBU$ cyclic
$\Leftrightarrow \angle AUB + \angle AWB = 180°$.

AC diameter
$\Leftrightarrow \angle ABC = 90°$.

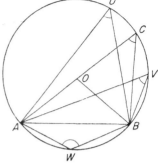

Fig. 9

Note. If $ABCD$ is a quadrilateral in which $\angle B = \angle D = 90°$, then $\angle A + \angle C = 180°$ and the quadrilateral is cyclic with AC as a diameter. This result is often important.

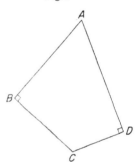

Fig. 10

(x) Tangency.

AT is tangent, AO is radius
$\Leftrightarrow OA \perp AT$
$\Leftrightarrow \angle TAP = \angle AUP$.

Fig. 11

(xi) SECANT THEOREM.

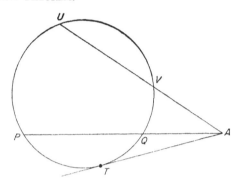

Fig. 12

P, Q, U, V concyclic
$\Leftrightarrow AP \cdot AQ = AU \cdot AV$.
AT is tangent at T to $\odot TPQ$
$\Leftrightarrow AT^2 = AP \cdot AQ$.

(xii) ANGLE BISECTOR THEOREMS.
$$\frac{BP}{PC} = \frac{BA}{AC} = \frac{BQ}{QC}$$
$\Leftrightarrow AP, AQ$ are the internal and external bisectors of $\angle A$.
Note that $PA \perp AQ$.

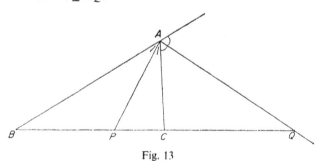

Fig. 13

(xiii) SOME LOCI.

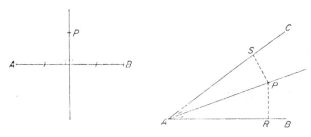

Fig. 14

(a) The locus of a point P equidistant from A, B is the perpendicular bisector of AB.

(b) The locus of a point P equidistant from two lines AB, AC is a bisector of $\angle BAC$.

TWO
The Geometry of the Triangle

1. The Centroid

THE PROBLEM. ABC is a given triangle and A', B', C' are the middle points of the sides BC, CA, AB.† It is required *to prove that AA', BB', CC' meet at a point G where each is trisected.*

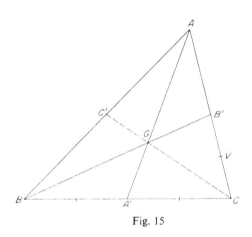

Fig. 15

Definitions. The lines AA', BB', CC' are called the *medians* of the triangle ABC.

The point G is called the *centroid* of the triangle ABC.

THE DISCUSSION. Let $G = AA' \cap BB'$, and let V be the middle point of CB'. Then

---

† In such contexts we omit the word "respectively", which remains understood.

$$CA' = A'B \brace CV = VB'} \Rightarrow A'V \| BB' \Rightarrow A'V \| GB'.$$

Also

$$GB' \| A'V \Rightarrow \frac{AG}{GA'} = \frac{AB'}{B'V}$$

$$= \frac{2}{1} \quad (AB' = B'C = 2B'V).$$

Thus $BB' \cap AA'$ is a point of trisection of AA'.

Similarly $CC' \cap AA'$ is the same point of trisection of AA'.

Hence AA', BB', CC' have a common point G which is a point of trisection of AA' and, by similar argument, of BB' and $CC.'$

Theorems

[Examples with which the reader should attempt to become familiar will be called *Theorems*.]

1. G is the centroid of $\triangle A'B'C'$.
2. If $P \in AB$, $Q \in AC$ such that $PQ \| BC$, and if $R = BQ \cap CP$ then $R \in AA'$.

Problems

1. $A'B'AC'$ is a parallelogram.
2. If AA' is produced to U so that $GU = AG$, then $UBGC$ is a parallelogram.
3. $B'C' \cap AG$ is a point of quadrisection of AG.
4. If $ABCD$ is a parallelogram, the centroids of $\triangle BAD$ and of $\triangle BCD$ are the points of trisection of AC.

Also the centroids of $\triangle ABC$, $\triangle BCD$, $\triangle CDA$, $\triangle DAB$ are at the vertices of a parallelogram.

5. If G, H are the centroids of two triangles ABC, ABD with a common side AB, then $GH \| CD$ and $GH = \frac{1}{3}CD$.

2. The Circumcentre

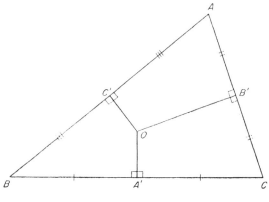

Fig. 16

THE PROBLEM. *ABC* is a given triangle. It is required *to establish the existence of a point O such that $OA = OB = OC$*; that is, $\exists O$ *such that* $OA = OB = OC$.

Definitions. The point O is called the *circumcentre* of $\triangle ABC$. The circle of centre O and radius OA passes through A, B, C and is called the *circumcircle* of $\triangle ABC$.

THE DISCUSSION. Let the perpendicular bisectors of AB, AC meet in O. Then

$O \in$ perpendicular bisector of $\begin{cases} AB \\ AC \end{cases}$ respectively

$\Rightarrow O$ is equidistant from $\begin{cases} A \text{ and } B \\ A \text{ and } C \end{cases}$ respectively

$\Rightarrow O$ is equidistant from A, B and C

$\Rightarrow OA = OB = OC$.

The Geometry of the Triangle 17

COROLLARIES. (i) $OA' \perp BC$, $OB' \perp CA$, $OC' \perp AB$. (ii) The perpendicular bisectors of the sides of a triangle are concurrent, in the circumcentre.

3. The Orthocentre

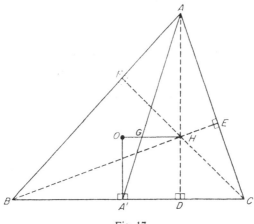

Fig. 17

THE PROBLEM. ABC is a given triangle, and $AD \perp BC$, $BE \perp CA$, $CF \perp AB$. It is required *to prove that AD, BE, CF have a common point H*.

Definitions. The lines AD, BE, CF are called the *altitudes* of $\triangle ABC$. The point H is called the *orthocentre* of $\triangle ABC$.

THE DISCUSSION. Let A' be the middle point of BC, so that the centroid G is the point on AA' such that $A'G/GA = 1/2$. Also let O be the circumcentre, so that $OA' \perp BC$.

Produce OG to H so that $OG/GH = 1/2$.
Then
$$OG/GH = 1/2 = A'G/GA$$
$$\Rightarrow \quad OA' \| AH.$$

But
$$OA' \perp BC,$$
so that
$$AH \perp BC.$$

Thus, the fixed point H on OG produced such that $OG/GH = 1/2$ has the property that $AH \perp BC$; hence AH is an altitude. Identical reasoning shows that BH, CH are altitudes. Hence the altitudes intersect in H.

COROLLARY. The points O, G, H are collinear and such that $OG/GH = 1/2$.

Definition. The line OGH is called the *Euler line* of $\triangle ABC$.

Theorems
1. O is the orthocentre of $\triangle A'B'C'$.
2. The four points A, B, C, H are so related that each is the orthocentre of the triangle whose vertices are the other three.
3. H is outside $\triangle ABC \Leftrightarrow$ one of the angles of the triangle is obtuse.
4. $\odot HBC = \odot HCA = \odot HAB = \odot ABC$ (That is, the radii are equal).
5. $AH \cdot HD = BH \cdot HE = CH \cdot HF$.
6. DA, BC bisect the angles between DE, DF.

Problems
1. The circles on AB, AC as diameters meet on BC.
2. The middle point of OH is equidistant from A' and D.
3. The triangle UVW is drawn so that $A \in VW$, $B \in WU$, $C \in UV$ and $WV \parallel BC$, $UW \parallel CA$, $VU \parallel AB$. Prove that H is the circumcentre of $\triangle UVW$.
4. $ABCD$ is a parallelogram; O is the circumcentre of $\triangle ABC$ and Q is the circumcentre of $\triangle ADC$. Prove that $AOCQ$ is a rhombus.
5. $ABCD$ is a parallelogram and H is the orthocentre of $\triangle ABC$. Prove that $D \in \odot AHC$.
6. The tangent at A to $\odot AEF$ is parallel to BC (Fig. 17).
7. The tangent at A to $\odot ABC$ is parallel to EF (Fig. 17).
8. (Alternative proof of the orthocentre property). In $\triangle ABC$, let BE, CF be altitudes meeting in H, and let AH meet BC in K. Establish the argument:
$$\angle KAE = \angle HFE = \angle KBE \Rightarrow AK \perp BC.$$

4. The Incentre and the Escribed Centres
(i) The incentre

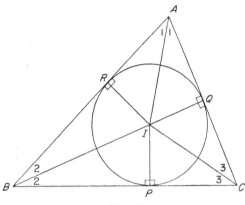

Fig. 18

THE PROBLEM. *ABC* is a given triangle. It is required *to establish the existence of a point I which lies inside the triangle and which is equidistant from BC, CA and AB;* that is, $\exists I$ such that, if $IP \perp BC$, $IQ \perp CA$, $IR \perp AB$, then $IP = IQ = IR$.

Definitions. The point *I* is called the *incentre* of $\triangle ABC$. The circle of centre *I* and radius *IP* is called the *incircle* of $\triangle ABC$; the lines *BC*, *CA*, *AB* are the tangents at *P*, *Q*, *R*.

THE DISCUSSION. Let the internal bisectors of $\angle B$, $\angle C$ meet in *I*. Then

$I \in$ bisector of $\begin{cases} \angle B \\ \angle C \end{cases}$ respectively

$\Rightarrow I$ is equidistant from the lines $\begin{cases} BA \text{ and } BC \\ CA \text{ and } CB \end{cases}$ respectively

$\Rightarrow I$ is equidistant from *BC*, *CA*, *AB*

20 Deductive Geometry

COROLLARY. The internal bisectors of the angles of a triangle are concurrent, in the incentre.

(ii) The escribed centres

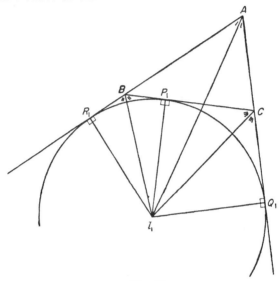

Fig. 19

THE PROBLEM. There are three circles to be described, one opposite each of the vertices A, B, C of the given triangle; the circle opposite A is taken as typical.

It is required *to establish the existence of a point I_1, lying outside the triangle but within the angle formed by AB and AC, such that I_1 is equidistant from BC, CA, AB;* that is, $\exists I_1$ such that, if $I_1P_1 \perp BC$, $I_1Q_1 \perp CA$, $I_1R_1 \perp AB$, then $I_1P_1 = I_1Q_1 = I_1R_1$.

Definitions. The point I_1 is called *the escribed centre opposite A* of $\triangle ABC$.

The circle of centre I_1 and radius I_1P_1 is called *the escribed circle opposite A* of $\triangle ABC$; the lines BC, CA, AB are the tangents at P_1, Q_1, R_1.

The escribed centres opposite B and C, and the escribed circles opposite B and C, are defined similarly.

THE DISCUSSION. Let the *external* bisectors of $\angle B$, $\angle C$ meet in I_1. Then

$I_1 \in$ a bisector of $\begin{cases} \angle B \\ \angle C \end{cases}$ respectively

$\Rightarrow I_1$ is equidistant from the lines $\begin{cases} BA \text{ and } BC \\ CA \text{ and } CB \end{cases}$ respectively

$\Rightarrow I_1$ is equidistant from BC, CA, AB.

(iii) The configuration of these four centres

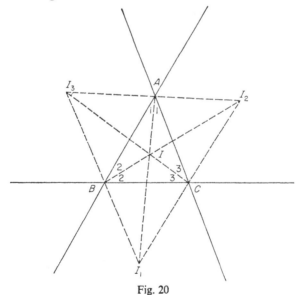

Fig. 20

22 *Deductive Geometry*

The diagram shows how the four points I, I_1, I_2, I_3 are related to the vertices A, B, C of the given triangle. The properties which follow are important, but the proofs are left to the reader.

Theorems

1. The points A, I, I_1 are collinear.
2. The points I_2, A, I_3 are collinear.
3. $IA \perp I_2I_3$; $IB \perp I_3I_1$; $IC \perp I_1I_2$.
4. I is the orthocentre of $\triangle I_1I_2I_3$.
5. If $BC = a$, $CA = b$, $AB = c$, $s = \tfrac{1}{2}(a + b + c)$, then $AQ = AR = s - a$, $AQ_1 = AR_1 = s$, $BP_1 = BR_1 = s - c$.
6. If $\triangle$ is the area of the triangle ABC, and if the radii of the inscribed and escribed circles are r, r_1, r_2, r_3, then

$$r = \triangle/s, r_1 = \triangle/(s - a).$$

Problems

1. $I_1 \in \odot BIC$.
2. $AI \cdot II_1 = BI \cdot II_2 = CI \cdot II_3$.
3. $BC = CA = AB \Leftrightarrow I_2I_3 = I_3I_1 = I_1I_2$.

5. The Nine-Points Circle

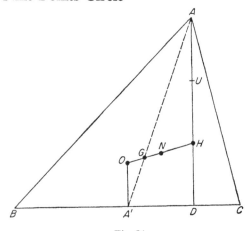

Fig. 21

The Geometry of the Triangle

THE PROBLEM. Let ABC be a given triangle, A', B', C' the middle points of BC, CA, AB; $AD \perp BC$, $BE \perp CA$, $CF \perp AB$; U, V, W the middle points of HA, HB, HC.

(Only A', D, U are shown in the diagram.)

It is required *to prove that the nine points A', B', C', D, E, F, U, V, W lie on a circle whose centre is N, the middle point of OH, and whose radius is one-half of that of $\odot ABC$.*

Definition. The circle is called the *nine-points circle* and N the *nine-points centre* of $\triangle ABC$.

THE DISCUSSION.

$\left. \begin{array}{l} A'O \| HA \\ A'G = \tfrac{1}{2} GA \end{array} \right\} \Rightarrow OA' = \tfrac{1}{2} HA = HU = UA.$

$\left. \begin{array}{l} A'O \| UA \\ A'O = UA \end{array} \right\} \Rightarrow A'O\,AU$ is a parallelogram

$\qquad\qquad \Rightarrow A'U = OA =$ radius of $\odot ABC$.

$\left. \begin{array}{l} A'O \| HU \\ A'O = HU \end{array} \right\} \Rightarrow A'OUH$ is a parallelogram

$\qquad\qquad \Rightarrow A'U$ is bisected at N, the middle point of OH
$\qquad\qquad \Rightarrow NU = NA' = \tfrac{1}{2}$ radius of $\odot ABC$.

Also

$\left. \begin{array}{l} OA' \perp BC,\ HD \perp BC \\ N \text{ is the middle point of } OH \end{array} \right\} \Rightarrow N$ is on the perpendicular bisector of $A'D$

$\qquad\qquad\qquad\qquad\qquad \Rightarrow ND = NA' = \tfrac{1}{2}$ radius of $\odot ABC$.

Hence
$$NA = NU = ND = \tfrac{1}{2} \text{ radius of } \odot ABC.$$

Similar argument obtains the same values for NB, NV, NE; NC, NW, NF.

Hence the nine points lie on a circle of centre N and radius equal to $\tfrac{1}{2}$ radius of $\odot ABC$.

6. The Nine-Points Circle; Alternative Treatment

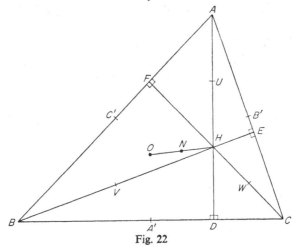

Fig. 22

THE PROBLEM. Let ABC be a given triangle, with circumcentre O and orthocentre H. Let A', B', C' be the middle points of BC, CA, AB; and $AD \perp BC$, $BE \perp CA$, $CF \perp AB$, so that AD, BE, CF meet in the orthocentre H; finally, let U, V, W be the middle points of HA, HB, HC and N the middle point of OH.

It is required *to prove that the points A', B', C', D, E, F, U, V, W all lie on a circle of centre N*.

THE DISCUSSION.

$$\left.\begin{array}{l} BA' = A'C \\ BV = VH \end{array}\right\} \Rightarrow A'V \parallel CH$$
$$\left.\begin{array}{l} CA' = A'B \\ CW = WH \end{array}\right\} \Rightarrow A'W \parallel BH$$
$$\Bigg\} \Rightarrow \angle VA'W = \angle VHW$$

But

$$\angle VHW = \angle EHF$$
$$= 180° - \angle EAF \quad (HF \perp AF \text{ and } HE \perp AE)$$
$$= 180° - \angle BAC$$

Also
$$\left.\begin{array}{l}HV = VB \\ HU = UA\end{array}\right\} \Rightarrow VU \| BA$$
$$\left.\begin{array}{l}HW = WC \\ HU = UA\end{array}\right\} \Rightarrow WU \| CA$$
$$\Rightarrow \angle VUW = \angle BAC.$$

Thus
$$\angle VA'W + \angle VUW = 180°.$$

Hence $VUWA'$ is cyclic; that is,
$$A' \in \odot UVW.$$

Similarly B', $C' \in \odot UVW$.
Again
$$\left.\begin{array}{l}\angle BDH = 90° \\ V \text{ is middle point of } BH\end{array}\right\} \Rightarrow VH = VD$$
$$\Rightarrow \angle VDH = \angle VHD.$$

Similarly,
$$\angle WDH = \angle WHD.$$

Hence
$$\begin{aligned}\angle VDW &= VDH + \angle WDH \\ &= \angle VHD + \angle WHD \\ &= \angle VHW \\ &= \angle VA'W \quad \text{as before.}\end{aligned}$$

Thus $VA'DW$ is cyclic; that is,
$$D \in \odot VA'W,$$
so that, as above,
$$D \in \odot UVWA'B'C'.$$

Similarly E, $F \in$ that circle. Hence
$$U, V, W, \quad A', B', C', \quad D, E, F$$
lie on a circle.

Finally,

$$\left.\begin{array}{l}HN = \tfrac{1}{2} HO\\ HU = \tfrac{1}{2} HA\end{array}\right\} \Rightarrow NU = \tfrac{1}{2} OA\text{; that is,}$$

$$NU = \tfrac{1}{2} \text{(radius of } \odot ABC\text{)}.$$

Similar NV, $NW = \tfrac{1}{2}$ (radius of $\odot ABC$).

Hence $NU = NV = NW$, so that N is the centre of the circle UVW.

Theorems

1. O, G, N, H are collinear, and $OG/GN = OH/HN = 2/1$.
Definition. The line $OGNH$ is called the *Euler line* of $\triangle ABC$.
2. A, B, C, H are the inscribed and escribed centres of $\triangle DEF$.
3. The triangles ABC, HBC, HCA, HAB all have the same nine-points circle.

Problems

1. $A'U$, $B'V$, $C'W$ bisect each other at N.
2. $\triangle UVW \equiv \triangle A'B'C'$.
3. $B'C'VW$ is a rectangle.
4. $UE = UF$; $A'U \perp EF$.
5. OU bisects AA'; $OU \cap AA' \in B'C'$.
6. UE, UF are the tangents from U to $\odot BCEF$.
7. If HD is produced to D' and HA' produced to A'' so that $HD = DD'$, $HA' = A'A''$, then D', A'' both lie on $\odot ABC$.
8. If $\odot ABC$ meets $\odot AFHE$ again in K, then the line OU is the perpendicular bisector of AK.

If $L = AK \cap BC$, then $L \in \odot EKC$.

By applying the theorem that the common chords of three circles taken in pairs are concurrent, or otherwise, prove that $L = EF \cap BC$.

THREE
Some Circle Theorems

1. The Theorem of Ptolemy

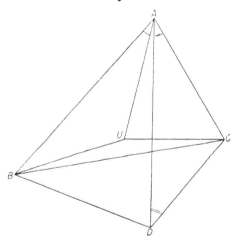

Fig. 23

(i) A quadrilateral inequality

THE PROBLEM. ABC is a given triangle and D an arbitrary point. (The diagram shows a typical position of D; other cases are possible, but the differences are slight.) A triangle ABU is constructed as in the diagram so that

$$\triangle ABU \sim \triangle ADC.$$

27

We are *to prove that*

$$\triangle ACU \sim \triangle ADB$$

and hence that

$$AB \cdot CD + AC \cdot BD > BC \cdot AD.$$

THE DISCUSSION.

$$\begin{aligned}\triangle ABU \sim \triangle ADC &\Rightarrow \angle BAU = \angle DAC \\ &\Rightarrow \angle BAU + \angle UAD = \angle UAD + \angle DAC \\ &\Rightarrow \angle BAD = \angle UAC.\end{aligned}$$

Also

$$\begin{aligned}\triangle ABU \sim \triangle ADC &\Rightarrow \frac{AB}{AD} = \frac{AU}{AC} \\ &\Rightarrow \frac{AB}{AU} = \frac{AD}{AC}.\end{aligned}$$

Hence, in $\triangle ACU$, $\triangle ADB$,

$$\angle UAC = \angle BAD$$

and

$$\frac{AU}{AB} = \frac{AC}{AD},$$

so that

$$\triangle ACU \sim ADB.$$

Moreover, from the similar triangles,

$$\begin{aligned}\triangle ABU \sim \triangle ADC &\Rightarrow \frac{AB}{AD} = \frac{BU}{DC} \\ &\Rightarrow AB \cdot CD = AD \cdot BU\end{aligned}$$

and

$$\begin{aligned}\triangle ACU \sim \triangle ADB &\Rightarrow \frac{AC}{AD} = \frac{CU}{BD} \\ &\Rightarrow AC \cdot BD = AD \cdot CU\end{aligned}$$

Some Circle Theorems

Hence

$$AB \cdot CD + AC \cdot BD = AD\,(BU + CU) > AD \cdot BC$$
(triangle inequality)

(ii) The cyclic case; Ptolemy's theorem.

The final step of the preceding work,

$$BU + CU > BC,$$

presupposed that U was not on BC, a condition that would certainly hold in general. The case of exception must now receive attention.

THE PROBLEM. Suppose that, in the preceding work, $U \in BC$. Then

$$BU + UC = BC$$

so that

$$AB \cdot CD + AC \cdot BD = AD \cdot BC.$$

It is required *to prove that*

$$AB \cdot CD + AC \cdot BD = AD \cdot BC$$
$$\Leftrightarrow A, B, C, D \text{ are concyclic}.$$

[Note the double arrow $\Leftrightarrow$.]

Suppose, first, that equality holds, so that $U \in BC$. Then

$$\angle ADC = \angle ABU$$
$$= \angle ABC \;\; (BUC \text{ is a straight line})$$

Hence A, B, C, D are concyclic.

Suppose, next, that A, B, C, D are concyclic. Then

$$\angle ABU = \angle ADC \text{ (given, since } \triangle ABU \sim \triangle ADC)$$
$$= \angle ABC \text{ (same segment)}$$

so that BU, BC are the same lines.

Hence $U \in BC$, so that the equality holds.

Note. The four points A, B, C, D can be split in two pairs in three ways,

$$AB, CD; \ AC, BD; \ BC, AD.$$

Ptolemy's theorem asserts that, when the four points are concyclic, the sum of products from two of these pairs is equal to the third. The pair which comes "third" is that defined by the diagonals of the cyclic quadrilateral.

[For a more detailed discussion of implications, see E. A. Maxwell, *Fallacies in Mathematics*, Cambridge University Press (1959) p. 28.]

Problems
1. If D is a point on the arc opposite A of the circumcircle of an equilateral triangle ABC, then $AD = BD + CD$.
2. AB, PQ are parallel chords of a circle. Prove that

$$AB \cdot PQ = AQ^2 - AP^2.$$

3. Identify the well-known result which is a special case of the theorem of Ptolemy when $ABCD$ is a rectangle.
4. In $\triangle ABC$, draw $BE \perp AC$, $CF \perp AB$. Prove that $BF = BC \cos B$, $CF = BC \sin B$, and write down similar expressions for CE, BE.

By applying the theorem of Ptolemy to $\odot BCEF$, prove that $EF = BC \cos A$.

5. AB is a diameter of a circle; P, Q are points on the circle, one in each half, so that $\angle PAB = \theta$, $\angle QAB = \phi$. Use the theorem of Ptolemy to prove that $\sin \theta \cos \phi + \sin \phi \cos \theta = \sin (\theta + \phi)$.

2. The Simson Line

(*Author's note.* It is possible to argue that the Simson line theorem as we know it is slightly unfortunate. Associated with a point P on the circumcircle of a triangle ABC are *several* parallel lines, of which the most significant is surely the one through

Some Circle Theorems

the orthocentre *H*. For the benefit of pupils taking examinations, the traditional proof is given first. A more general discussion then follows, based on the line to which reference has just been made.)

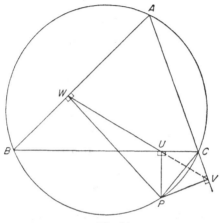

Fig. 24

THE PROBLEM. Take an arbitrary point *P* on the circumcircle of a triangle *ABC*. Draw $PU \perp BC$, $PV \perp CA$, $PW \perp AB$. It is required *to prove that U, V, W are collinear.*

Definition. *UVW* is known as the *Simson line of P* with respect to the triangle *ABC*.

THE DISCUSSION.

$$\left.\begin{array}{l} PU \perp BC \\ PV \perp CA \end{array}\right\} \Rightarrow PUCV \text{ cyclic} \Rightarrow \angle CUV = \angle CPV.$$

$$\left.\begin{array}{l} PU \perp BC \\ PW \perp AB \end{array}\right\} \Rightarrow PUWB \text{ cyclic} \Rightarrow \angle BUW = \angle BPW.$$

Also

$$\left.\begin{array}{l} PV \perp CA \\ PW \perp AB \end{array}\right\} \Rightarrow PVAW \text{ cyclic} \Rightarrow \angle VPW = 180° - A$$

and
$$ABPC \text{ cyclic} \Rightarrow \angle BPC = 180° - A.$$
Hence
$$\angle VPW = \angle BPC,$$
so that, subtracting $\angle WPC$ from each,
$$\angle CPV = \angle BPW.$$
Hence
$$\angle CUV = \angle BUW,$$
so that
$$UVW \text{ is a straight line.}$$

Note. The *converse theorem* is also true:

Given a point P in the plane of a triangle ABC, and $PU \perp BC$, $PV \perp CA$, $PW \perp AB$, then, if U, V, W, are collinear, $P \in \odot ABC$.

The proof is very nearly a reversal of the steps just given.

3. The Simson Property; Alternative Treatment

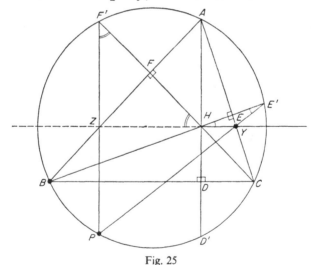

Fig. 25

(i) Lemma

Consider first a theorem which is important in its own right:

THE PROBLEM. The altitudes AHD, BHE, CHF of a triangle ABC meet $\odot ABC$ in D', E', F'. It is required *to prove that D', E', F' are the mirror images of H in BC, CA, AB*. Compare Problem 7 on p. 26.

[By saying that D' is the *mirror image* of H in BC we mean that, if $D = HD' \cap BC$, then $HD' \perp BC$ and $HD = D'D$.]

THE DISCUSSION.
$$\left.\begin{array}{r}HD \perp BC \\ HE \perp CA\end{array}\right\} \Rightarrow HDCE \text{ cyclic} \Rightarrow \angle BHD = \angle ACB.$$

But
$$\angle ACB = \angle AD'B \text{ (same segment)},$$
so that
$$\angle BHD = \angle BD'D \text{ (re-naming the latter angle)}.$$

Hence, in $\triangle BHD$, $BD'D$,
$$BD = BD;\ \angle BDH = \angle BDD',\ \angle BHD = \angle BD'D,$$
so that
$$\triangle BHD \equiv \triangle BD'D.$$

In particular,
$$HD = D'D,$$
so that D' is the mirror image of H in BC.

Similarly for E', F'.

(ii) The main theorem

THE PROBLEM. Let P be an arbitrary point on $\odot ABC$. With the notation of the Lemma, let

$$X = PD' \cap BC,\ Y = PE' \cap CA,\ Z = PF' \cap AB.$$

[The point X is not shown in the diagram.]

It is required *to prove that X, Y, Z lie on a straight line which passes through H*.

THE DISCUSSION. It is an immediate consequence of the preceding work that

$$\angle YHE' = \angle YE'H$$
$$= \angle PE'B \quad \text{(re-naming)}$$
$$= \angle PAB \quad \text{(same segment)}$$

and that

$$\angle ZHF' = \angle ZF'H$$
$$= \angle PF'C \quad \text{(re-naming)}$$
$$= \angle PAC \quad \text{(same segment)}$$

Also

$$\angle EHF = 180° - \angle BAC \quad (HE \perp CA, HF \perp AB).$$

Adding corresponding sides of these three equations,

$$\angle YHE' + \angle EHF + \angle ZHF' = 180° + \angle PAB + \angle PAC - \angle BAC = 180°.$$

Hence YHZ is a straight line.

That, is, $Y \in$ line ZH; and similar argument shows that $X \in$ line ZH. Hence X, Y, Z, H are collinear.

Remark: The pith of this proof is the simple observation, "angle on arc BP + angle on arc PC = angle on arc BC."

(iii) Some consequences

The diagram is now drawn with special reference to the vertex A.

THE PROBLEM. With notation as before, let $PU \perp BC$, and let the line PU meet $\odot ABC$ again in U'. We prove first that *the line AU' is parallel to $XYZH$*; next that, *if Q is the middle point of PH, then UQ is parallel to $XYZH$*; and finally, as an immediate consequence, that *if V, W are defined similarly to U, then U, V, W all line on the line through Q parallel to $XYZH$*.

Some Circle Theorems

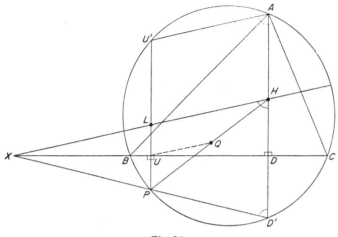

Fig. 26

THE DISCUSSION. By the preceeding work,

$$\angle XHD = \angle XD'D$$
$$= \angle PD'A \quad \text{(re-naming)}$$
$$= 180° - \angle PU'A \quad (\odot AD'PU')$$
$$= \angle U'AH \quad (PU'\|D'A).$$

But these are corresponding angles, and so

$$HX\|AU'.$$

Next, let PU meet $XYZH$ in L. Thus, immediately, since $D'D = DH$, we have

$$PU = UL.$$

Then

$$\left.\begin{array}{r}PU = UL \\ PQ = QH\end{array}\right\} \Rightarrow UQ\|LH,$$

so that UQ is parallel to $XYZH$.

Finally, if $PV \perp CA$, $PW \perp AB$, then, similarly,

$$VQ \| XYZH, \ WQ \| XYZH.$$

Thus U, V, W all lie on a straight line (the *Simson line* of P with respect to the triangle ABC) which is the line through the middle point of PH and parallel to $XYZH$.

COROLLARY. The Simson line of P bisects HP.

Problems

1. $BHCP$ is a parallelogram $\Leftrightarrow$ AP is a diameter of $\odot ABC$.
2. PR is a diameter of $\odot ABC$ $\Leftrightarrow$ Simson line of P $\perp$ Simson line of R.
3. If K is the middle point of the arc BC remote from A, then the Simson line of K is the line through the middle point of BC perpendicular to AK.
4. If PR is a chord of $\odot ABC$ perpendicular to BC, then the (supplementary) angles between the Simson lines of P and R are equal to the angles subtended at the circumference by PR.
5. If the altitude AD of $\triangle ABC$ meets $\odot ABC$ again in P, then the Simson line of P is parallel to the tangent at A.
6. A straight line cuts the sides BC, CA, AB of a triangle ABC in points L, M, N; the circles ABC, AMN meet in a further point P. Prove that the feet of the perpendiculars from P to BC, CA, AB, LMN are collinear, and deduce that $P \in \odot BNL$, $P \in \odot CLM$.

4. Centres of Similitude

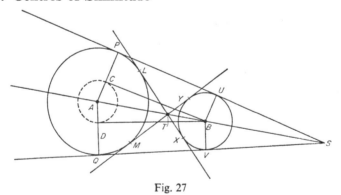

Fig. 27

Some Circle Theorems

THE PROBLEM. Let two given circles have centres A, B and radii a, b, and suppose that their centres are distant d apart. We are to demonstrate precisely, what is, indeed, obvious intuitively, that the circles have two *direct* common tangents PU, QV: that is, lines which are tangents to both circles simultaneously; and (when, as in the diagram, the two circles lie entirely outside each other) two *transverse* common tangents LX, MY. It is also "obvious" that PU, QV meeet at a point S on AB and that LX, MY meet at a point T on AB; further, that AB bisects each of the angles $\angle PSQ$, $\angle LTM$.

THE DISCUSSION.

(i) *To construct PU, QV.* Draw the circle of centre A and radius $a - b$ (supposing that a is greater than b); construct the tangents BC, BD from B to this circle by the standard method of drawing the circle on AB as diameter to meet it in C, D; let AC, AD meet the circle of centre A in P, Q; let lines through B parallel to AP, AQ meet the circle of centre B in U, V. Then PU, QV are the direct common tangents.

To prove this:

$$\left.\begin{array}{l} AC = a - b \\ AP = a \end{array}\right\} \Rightarrow CP = b = BU$$

$$\left.\begin{array}{l} CP = BU \\ CP \| BU \end{array}\right\} \Rightarrow CPUB \text{ is a parallelogram}$$

C on circle of diameter $AB \Rightarrow \angle ACB = 90° \Rightarrow \angle PCB = 90°$
$\Rightarrow CPUB$ is a rectangle
$\Rightarrow PU \perp AP$ and BU
$\Rightarrow \begin{cases} PU \text{ is tangent at } P \\ UP \text{ is tangent at } U. \end{cases}$

Similarly QV is tangent at Q and at V.

The construction for LX, MY is very similar, save that the first step is to draw the circle of centre A and radius $a + b$.

(ii) *To locate S and T.* Let PU meet AB in S. Then
$$BU \| AP \Rightarrow \triangle SBU \sim \triangle SAP$$
$$\Rightarrow \frac{SB}{SA} = \frac{BU}{AP} = \frac{b}{a}.$$

Thus
$$\frac{SA}{a} = \frac{SB}{b} = \frac{SA - SB}{a - b} = \frac{d}{a - b},$$

so that
$$SA = \frac{ad}{a - b}, \quad SB = \frac{bd}{a - b}.$$

Identical argument shows that QV meets AB in precisely the same point.

In the same way, LX, MY meet AB in the point T where
$$BX \| AL \Rightarrow \triangle TBX \sim \triangle TAL$$
$$\Rightarrow \frac{TB}{TA} = \frac{BX}{AL} = \frac{b}{a},$$

so that
$$\frac{TA}{a} = \frac{TB}{b} = \frac{TA + TB}{a + b} = \frac{d}{a + b}$$

giving
$$TA = \frac{ad}{a + b}, \quad TB = \frac{bd}{a + b}.$$

Note, incidentally, the ratios
$$\frac{SB}{SA} = \frac{TB}{TA} = \frac{b}{a}.$$

(iii) *The lengths of PU, LX.* Using the theorem of Pythagoras, we have
$$PU^2 = CB^2 = AB^2 - AC^2 = d^2 - (a - b)^2$$
$$= (d + a - b)(d - a + b)$$

and, similarly,
$$LX^2 = d^2 - (a + b)^2 = (d + a + b)(d - a - b).$$

(iv) *Similitude*

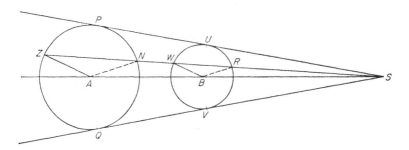

Fig. 28

Definition. The points S, T are called the *centres of similitude* of the two given circles.

The name *similitude* is justified as follows:

Let AZ, BW be parallel radii (in the same sense) of the two circles. We prove that *the line ZW passes through S*. For $AZ \| BW$
$\Rightarrow \angle SAZ = \angle SBW$

and

$$\frac{AZ}{BW} = \frac{a}{b} = \frac{SA}{SB} \Rightarrow \frac{SA}{AZ} = \frac{SB}{BW},$$

so that

$$\triangle SAZ \sim \triangle SBW.$$

In particular,

$$\angle ASZ = \angle BSW,$$

so that SWZ is a straight line.

To follow the configuration further, let the line SWZ cut the circles again in N, R. Then *we prove that $AN \| BR$*:

$$AN = AZ \Rightarrow \angle ANZ = \angle AZN,$$
$$AZ \| BW \Rightarrow \angle AZN = \angle BWR,$$
$$BW = BR \Rightarrow \angle BWR = \angle BRW,$$

so that
$$\angle ANZ = \angle BRW,$$
giving
$$AN \| BR.$$

The two circles are thus, in an obvious sense of the phrase, *similarly placed* with respect to S and (by the same kind of argument) with respect to T.

Theorems
1. *More generally* two circles, however placed, have associated with them two points S, T on their line of centres such that the lines joining ends of parallel radii, one of each circle, pass through one or other of S, T.

The points, S, T are *centres of similitude* for the circles.

2. The orthocentre H and the centroid G are the centres of similitude for the nine-points circle and the circumcircle of a triangle ABC.

Problems
1. The centres of similitude for the incircle and the escribed circle opposite A of $\triangle ABC$ are the vertex A and the point where the internal bisector of $\angle BAC$ meets BC.
2. Two equal circles have only one centre of similitude.

5. Radical Axes

THE PROBLEM. Let two non-intersecting circles be given, of centres A, B and radii a, b. It is required to *prove that the locus of a point P such that the tangents from P to the two circles are equal is a straight line perpendicular to AB*.

Definition. The line is called the *radical axis* of the two circles.

THE DISCUSSION. Let P be such a point, and draw the tangents PT, PU to the circles. Then $PT = PU$.

But

PT is a tangent $\Rightarrow \angle ATP = 90°$
$\Rightarrow PT^2 = PA^2 - AT^2 = PA^2 - a^2,$

Some Circle Theorems

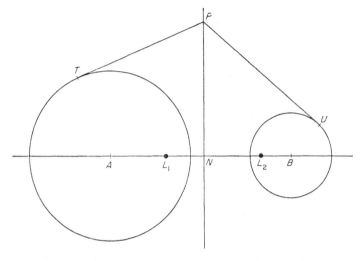

Fig. 29

and, similarly,
$$PU^2 = PB^2 - b^2.$$

Hence
$$PA^2 - a^2 = PB^2 - b^2,$$

or
$$PA^2 - PB^2 = a^2 - b^2.$$

Draw $PN \perp AB$.

Then
$$PA^2 - PB^2 = a^2 - b^2$$
$$\Rightarrow (PN^2 + AN^2) - (PN^2 + BN^2) = a^2 - b^2$$
$$\Rightarrow AN^2 - BN^2 = a^2 - b^2$$
$$\Rightarrow (AN + BN)(AN - BN) = a^2 - b^2.$$

Thus, if $AB = d$,
$$AN - BN = \frac{a^2 - b^2}{d}.$$

Also
$$AN + BN = d,$$
so that
$$AN = \frac{d^2 + a^2 - b^2}{2d},$$
$$BN = \frac{d^2 - a^2 + b^2}{2d}.$$

Hence N is a fixed point, and so *the point P lies on the fixed line through N perpendicular to AB.*

Further immediate properties.

Let the circle of centre P and radius PT cut the line AB in points L_1, L_2. We prove that L_1, L_2 *are the same for all positions of P on the radical axis.*

Definition. L_1, L_2 are called the *limiting points* of the two circles.

By the theorem of Pythagoras,

$$\begin{aligned} NL_1^2 &= PL_1^2 - PN^2 \\ &= PT^2 - (PA^2 - AN^2) \\ &= AN^2 - (PA^2 - PT^2) \\ &= AN^2 - AT^2 \\ &= \left(\frac{d^2 + a^2 - b^2}{2d}\right)^2 - a^2 \\ &= \left(\frac{d^2 + a^2 - b^2}{2d} + a\right)\left(\frac{d^2 + a^2 - b^2}{2d} - a\right) \\ &= \left\{\frac{(d+a)^2 - b^2}{2d}\right\}\left\{\frac{(d-a)^2 - b^2}{2d}\right\} \\ &= \frac{(d+a+b)(d+a-b)(d-a+b)(d-a-b)}{4d^2}. \end{aligned}$$

Similar argument obtains the same value for NL_2^2. Also, the expression on the right-hand side depends only on a, b, d, and so is independent of the position of P on the radical axis.

Some Circle Theorems

Hence L_1, L_2 are fixed points, and also $NL_1 = NL_2$.

Remark: The step "$NL_1^2 = AN^2 - AT^2$" suffices to prove that the position of L_1 is independent of P. The algebraic formulation is added for interest.

Theorems, (i) For intersecting circles

1. If two circles intersect in X, Y, then all points P, such that the tangents from P to the two circles are equal, lie on XY.
2. In Question 1, the circle of centre P and radius $\sqrt{(PX \cdot PY)}$ does not meet the line of centres of the two given circles.
3. The common chords of three circles taken in pairs are concurrent.

Theorems, (ii) For non-intersecting circles

1. The radical axes of three non-intersecting circles taken in pairs (with non-collinear centres) are concurrent.
2. If a circle cuts the circle of centre A in R, S and the circle of centre B in R', S', then $RS \cap R'S'$ is on the radical axis of the two given circles.

Theorems (Generalization)

Let P be a point in the plane of a given circle, and let an arbitrary line through P meet the circle in A, B. Then the product $\overrightarrow{PA} \cdot \overrightarrow{PB}$ (having regard to sign; compare p. 51) is called the *power* of P with respect to the circle. It is independent of the chord selected.

1. The power of P is positive, negative or zero according as P is outside, inside or on the circle.
2. The locus of a point P such that the powers of P with respect to the two circles are equal is a straight line, the *radical axis* of the circles.

Problems

1. In $\triangle ABC$, $AD \perp BC$, $BE \perp CA$, $CF \perp AB$. Verify that the common chords of $\odot ABC$, $\odot BCEF$ is BC, that the common chord of $\odot AEF$, $\odot BCEF$ is EF, and deduce that, if $X = BC \cap EF$ and if AX meets $\odot ABC$ again in Y, then $Y \in \odot AEF$.
2. The radical axis of the inscribed circle and the escribed circle opposite A of $\triangle ABC$ bisects BC.
3. In $\triangle ABC$, $AD \perp BC$, $BE \perp CA$, $CF \perp AB$; $P = BC \cap EF$, $Q = CA \cap FD$, $R = AB \cap DE$. Prove that $PE \cdot PF = PB \cdot PC$, and deduce that P, Q, R lie on a straight line perpendicular to the Euler line $OGNH$ of $\triangle ABC$.

4. Prove that the locus of a point P, which moves so that the length of the tangent from P to a given circle is equal to the distance from P to a given point A outside that circle, is a straight line.

5. In a copy of Fig. 29, everything was obliterated except the three points L_1, L_2, U. Starting from these three points as sole data, show how to construct the circle through U.

6. Orthogonal Circles

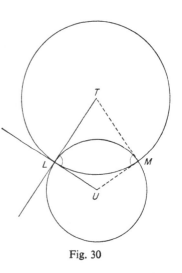

Preliminary definition: When two circles pass through a point L, *the angle between the circles* is defined to be the angle between the two tangents at L.

The two circles have a second common point M. It is clear from the symmetry of the figure, or easy to prove directly (in the diagram, $TL = TM$, $UL = UM \Rightarrow \triangle LTU \equiv \triangle MTU$), that the angle between the tangents at L is equal to the angle between the tangents at M; that is, *the angle between the circles is the same at each of their common points.*

Fig. 30

In particular,

Definition: Two circles are said to be *orthogonal* when they cut at right angles.

THE PROBLEM. Suppose that two orthogonal circles, of centres A, B, cut at L, M. We establish *two tests for orthogonality*:

(i) *the circles are orthogonal at $L(M)$ if the tangent at $L(M)$ to either passes through the centre of the other;*

Some Circle Theorems

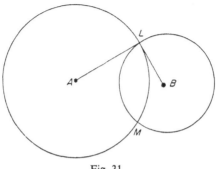

Fig. 31

(ii) *the circles are orthogonal if the square on the distance between their centres is equal to the sum of the squares on their radii.*

THE DISCUSSION (i) Taking the intersection at L as typical, draw the tangent there to the circle of centre A (B). If the circles are orthogonal, the tangent at L to the other circle is perpendicular to it and so passes through the centre of $A(B)$.

(ii) By the theorem of Pythagoras,

$$AL \perp LB \Rightarrow AB^2 = AL^2 + LB^2.$$

Notation: We sometimes write

$$\odot A \perp \odot B$$

(or equivalent) to denote that the circles of centres A and B are orthogonal.

The two tests just established are reversible, for it may be proved directly that

$$\odot A \perp \odot B \Leftrightarrow \text{tangent to either passes through}$$
$$\text{centre of other}$$
$$\Leftrightarrow AB^2 = a^2 + b^2.$$

7. Inverse Points

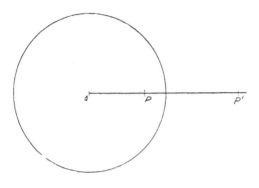

Fig. 32

Definition. Given a circle of centre A and radius a, two points P, P' are said to be *inverse with respect to it* if the points P, P' lie on a line through A, both on the same side of A, and if they are such that

$$AP \cdot AP' = a^2.$$

P' is called the *inverse* of P with respect to the circle; *then P is also the inverse of P'.*

Properties of circles through points inverse with respect to a given circle.

THE PROBLEM. Let P, P' be two points inverse with respect to a given circle of centre A. For convenience of reference, denote this circle by the symbol Ω (Greek *omega*).

We prove that (i) *every circle through P, P' cuts the circle Ω orthogonally*; also that (ii) *any line through A cuts every such circle in two points which are inverse with respect to Ω.*

Some Circle Theorems

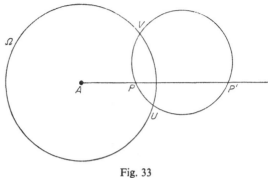

Fig. 33

THE DISCUSSION. (i) Let an arbitrary circle through P, P' cut Ω in points U, V. Then

P,P' inverse with respect to Ω
$\Rightarrow AU^2 = AP \cdot AP'$
$\Rightarrow AU$ is a tangent to the circle UPP'
$\Rightarrow$ the circles are orthogonal (since AU is a radius of Ω).

(ii) If an arbitrary line through A cuts the circle UPP' (which was drawn arbitrarily through P, P') in points L, M, not shown in the diagram, then

AU is a tangent to the circle UPP'
$\Rightarrow AL \cdot AM = AU^2$
$\Rightarrow L$, M are inverse with respect to Ω.

8. Further Properties of Radical Axes and Limiting Points

Given two circles of centres A, B and radii a, b, let RN be their radical axis, meeting AB in N, and let L_1, L_2 be the limiting points.

48   *Deductive Geometry*

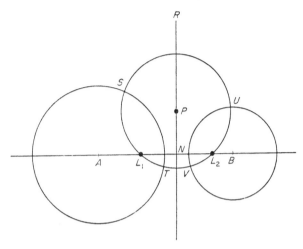

Fig. 34

THE PROBLEM. It is required *to prove that L_1, L_2 are inverse points with respect to each of the given circles.*

THE DISCUSSION. Take any point P on the radical axis and let the circle of centre P and radius PL_1 ($= PL_2$) cut the given circles in points S, T and U, V. Then, since (§5) PL_1 is the length of a tangent from P to either circle, PS, PT, PU, PV are tangents; so that

$$\odot P \perp \odot A; \quad \odot P \perp \odot B.$$

But

$$\left.\begin{array}{l}\odot P \perp \odot A \\ AL_1 L_2 \text{ cuts } \odot P \text{ in } L_1, L_2\end{array}\right\} \Rightarrow L_1, L_2 \text{ are inverse with respect to } \odot A.$$

Similarly L_1, L_2 are inverse with respect to $\odot B$.

COROLLARY. *Every circle through the limiting points cuts both the given circles orthogonally.*

Some Circle Theorems 49

Theorem

1. In $\triangle ABC$, $AD \perp BC$, $BE \perp CA$, $CF \perp AB$. Then $\exists$ a circle with respect to which B, F; C, E; H, D are three inverse pairs. This circle cuts orthogonally each of the circles $BCEF$, $BDHF$, $CDHE$.

Problems

1. In $\triangle ABC$, $BE \perp CA$, $CF \perp AB$. Prove that $\odot AEF \perp \odot BCEF$.

2. P is a point on the circle of diameter AB; prove that the circle with centre A and radius AP is orthogonal to the circle of centre B and radius BP.

Prove also that the circle of centre A and radius BP is orthogonal to the circle of centre B and radius AP.

9. Coaxal Circles

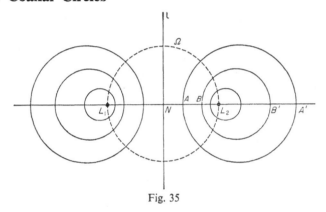

Fig. 35

Definition. L_1, L_2 are the extremities of a diameter of a given circle Ω of centre N. A number of pairs of points A, A'; B, B'; ... inverse with respect to Ω are taken on L_1L_2. The system of circles on AA', BB', ... as diameters is called a *coaxal system*, and L_1, L_2 are called the *limiting points of the system*.

THE PROBLEM. Let l be the perpendicular bisector of L_1L_2. The point of the name "coaxal system" is that *l is the radical axis of* any *two circles of the system*.

50   *Deductive Geometry*

THE DISCUSSION. Take, for example, the circles on AA', BB' as diameters. Then

A, A' and B, B' are inverse with respect to Ω

$\Rightarrow NA \cdot NA' = NB \cdot NB'$

$\Rightarrow$ the tangents from N to the circles are equal

$\Rightarrow N$ is on the radical axis of the circles

$\Rightarrow$ the radical axis is the line through N perpendicular to AB

$\Rightarrow$ the radical axis is l.

Theorems

1. The locus of a point P, which moves so that the ratio PA/PB of its distances from two fixed points A, B has constant value r, is a circle.

Definition. The circle is called the *circle of Apollonius* for the points A, B and the ratio r.

2. If, in Question 1, r is allowed to vary, then the circles for different values of r form a coaxal system with A, B as limiting points.

3. One and only one circle of a given coaxal system can be drawn through an arbitrary point U. Its centre is on the perpendicular bisector of UU', where U' is the inverse of U with respect to Ω.

4. *Every* circle through L_1, L_2 cuts orthogonally *every* circle of the coaxal system.

5. If two circles intersect at A, B, then (p. 43) their radical axis is the line AB. All pairs of circles through A, B have this same radical axis. Conversely, any circle whose radical axis when taken with the first circle is AB and whose radical axis when taken with the second circle is also AB passes through A and B.

6. (Compare Example 4 of this set). The circles through L_1, L_2 of Fig. 35 form a second coaxal system, of *intersecting circles*, and each circle of this system cuts each circle of the given system orthogonally.

FOUR
The Theorems of Ceva and Menelaus

1. The Idea of "Sense" on a Line

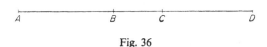

Fig. 36

There are many problems in which it is convenient to superpose on a line a *sense of description*, by which we mean that a symbol such as

$$\overrightarrow{AB}$$

denotes not only the length AB but also that it is to be regarded as traced by a point starting at A and moving so as to finish at B. It is then natural to adopt the algebraic symbolism

$$\overrightarrow{BA} = -\overrightarrow{AB}$$

for the line described in the opposite sense.

With this convention, *the equation*

$$\overrightarrow{AB} = \overrightarrow{AC} + \overrightarrow{CB}$$

is true for all collinear points A, B, C, whatever the order in which they occur. In particular,

$$\overrightarrow{AB} + \overrightarrow{BC} + \overrightarrow{CD} + \overrightarrow{DA} = 0.$$

2. The Theorem of Menelaus

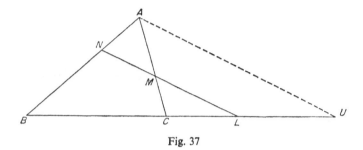

Fig. 37

THE PROBLEM. Let ABC be a given triangle, and suppose that any straight line (a *transversal*) is drawn to cut the sides BC, CA, AB in L, M, N. It is required *to prove that*

$$\frac{\overrightarrow{BL}}{\overrightarrow{LC}} \cdot \frac{\overrightarrow{CM}}{\overrightarrow{MA}} \cdot \frac{\overrightarrow{AN}}{\overrightarrow{NB}} = -1.$$

THE DISCUSSION. Let the line through A parallel to LMN cut BC in U. Then, *in whatever order the points occur along the lines*,

$$LM \| UA \Rightarrow \frac{\overrightarrow{CM}}{\overrightarrow{MA}} = \frac{\overrightarrow{CL}}{\overrightarrow{LU}},$$

$$LN \| UA \Rightarrow \frac{\overrightarrow{AN}}{\overrightarrow{NB}} = \frac{\overrightarrow{UL}}{\overrightarrow{LB}}.$$

Hence

$$\frac{\overrightarrow{BL}}{\overrightarrow{LC}} \cdot \frac{\overrightarrow{CM}}{\overrightarrow{MA}} \cdot \frac{\overrightarrow{AN}}{\overrightarrow{NB}} = \frac{\overrightarrow{BL}}{\overrightarrow{LC}} \cdot \frac{\overrightarrow{CL}}{\overrightarrow{LU}} \cdot \frac{\overrightarrow{UL}}{\overrightarrow{LB}} = -1.$$

3. The Theorem of Ceva

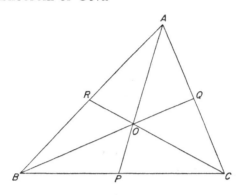

Fig. 38

THE PROBLEM. Let ABC be a given triangle, and O any point in its plane. Suppose that AO, BO, CO meet BC, CA, AB in P, Q, R. It is required *to prove that*

$$\frac{\vec{BP}}{\vec{PC}} \cdot \frac{\vec{CQ}}{\vec{QA}} \cdot \frac{\vec{AR}}{\vec{RB}} = +1.$$

THE DISCUSSION. Apply the theorem of Menelaus to the triangles APB, APC in turn.

COR is transversal of $\triangle APB$
$$\Rightarrow \frac{\vec{PC}}{\vec{CB}} \cdot \frac{\vec{BR}}{\vec{RA}} \cdot \frac{\vec{AO}}{\vec{OP}} = -1$$

and

BOQ is transversal of $\triangle APC$
$$\Rightarrow \frac{\vec{PB}}{\vec{BC}} \cdot \frac{\vec{CQ}}{\vec{QA}} \cdot \frac{\vec{AO}}{\vec{OP}} = -1.$$

Hence

$$\frac{\overrightarrow{PB}}{\overrightarrow{BC}} \cdot \frac{\overrightarrow{CQ}}{\overrightarrow{QA}} = -\frac{\overrightarrow{OP}}{\overrightarrow{AO}} = \frac{\overrightarrow{PC}}{\overrightarrow{CB}} \cdot \frac{\overrightarrow{BR}}{\overrightarrow{RA}},$$

so that

$$\frac{\overrightarrow{BP}}{\overrightarrow{PC}} \cdot \frac{\overrightarrow{CQ}}{\overrightarrow{QA}} \cdot \frac{\overrightarrow{AR}}{\overrightarrow{RB}} = +1.$$

4. Converse of the Theorems of Ceva and Menelaus

THE PROBLEM. The converses of the theorems of Ceva and Menelaus are also true.

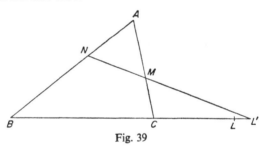

Fig. 39

(i) *Given points L, M, N on the sides BC, CA, AB of a triangle ABC such that*

$$\frac{\overrightarrow{BL}}{\overrightarrow{LC}} \cdot \frac{\overrightarrow{CM}}{\overrightarrow{MA}} \cdot \frac{\overrightarrow{AN}}{\overrightarrow{NB}} = -1,$$

to prove that L, M, N are collinear.

Let $L' = MN \cap BC$. Then, by the theorem of Menelaus,

$$\frac{\overrightarrow{BL'}}{\overrightarrow{L'C}} \cdot \frac{\overrightarrow{CM}}{\overrightarrow{MA}} \cdot \frac{\overrightarrow{AN}}{\overrightarrow{NB}} = -1,$$

so that
$$\frac{\vec{BL}}{\vec{LC}} = \frac{\vec{BL'}}{\vec{L'C}}.$$

Hence the points L, L' coincide, and so L lies on the line MN.

(ii) *Given points P, Q, R on the sides BC, CA, AB of a triangle ABC such that*
$$\frac{\vec{BP}}{\vec{PC}} \cdot \frac{\vec{CQ}}{\vec{QA}} \cdot \frac{\vec{AR}}{\vec{RB}} = +1,$$

to prove that AP, BQ, CR are concurrent.

Let $O = BQ \cap CR$, and $P' = AO \cap BC$. Then, by the theorem of Ceva,

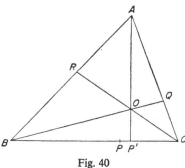

Fig. 40

$$\frac{\vec{BP'}}{\vec{P'C}} \cdot \frac{\vec{CQ}}{\vec{QA}} \cdot \frac{\vec{AR}}{\vec{RB}} = +1,$$

so that
$$\frac{\vec{BP}}{\vec{PC}} = \frac{\vec{BP'}}{\vec{P'C}}.$$

Hence the points P, P' coincide so that AP also passes through O.

Problems

1. Use the theorem of Ceva to establish for a given triangle the concurrence of (i) the medians, (ii) the altitudes, (iii) the internal bisectors of the angles. (But the proofs given earlier in this book are more revealing.)

2. Points D, E, F are taken on BC, CA, AB so that AD, BE, CF are concurrent, and $U = EF \cap BC$. Prove that

$$\vec{BD}/\vec{DC} = -\vec{BU}/\vec{UC}.$$

3. A point K is taken on the median AA' of $\triangle ABC$; $V = BK \cap AC$, $W = CK \cap AB$. Prove that $VW \parallel BC$.

4. The incircle of $\triangle ABC$ touches BC, CA, AB at P, Q, R. Prove that AP, BQ, CR are concurrent.

5. A point O is taken inside a triangle ABC; OA, OB, OC are produced to points P, Q, R; $L = BC \cap QR$, $M = CA \cap RP$, $N = AB \cap PQ$. Prove that

$$\frac{\vec{BL}}{\vec{LC}} \cdot \frac{\vec{CR}}{\vec{RO}} \cdot \frac{\vec{OQ}}{\vec{QB}} = -1,$$

and deduce that L, M, N are collinear. (The theorem of *Desargues*.)

FIVE
Harmonic Properties

1. The Definition

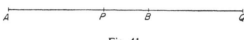

Fig. 41

Let AB be a given line on which there are also two points P, Q. The division of AB at the points P, Q defines two ratios $\vec{AP}/\vec{PB}$ and $\vec{AQ}/\vec{QB}$. There is a very large field of geometrical study in which the *ratio* of these two ratios is of vital importance:

Definition. The expression

$$\frac{\vec{AP}}{\vec{PB}} \bigg/ \frac{\vec{AQ}}{\vec{QB}}$$

is called the *cross-ratio* of the four collinear points.

In this book, however, we are not concerned with general values of the cross-ratio, but only with the particular case which arises when AB is divided in the same (numerical) ratio by P and by Q, internally and externally respectively, so that

$$\frac{\vec{AP}}{\vec{PB}} = -\frac{\vec{AQ}}{\vec{QB}}.$$

In this case, then, the value of the cross-ratio is -1.

Because of its great importance, a special name is required for this cross-ratio:

Definitions. A system of points on a straight line is called a *range*; a system of lines passing through a point is called a *pencil*. The line is called the *base* of the range and the point the *vertex* of the pencil.

When four points A, B, P, Q on a line form a range such that

$$\frac{\overrightarrow{AP}}{\overrightarrow{PB}} = -\frac{\overrightarrow{AQ}}{\overrightarrow{QB}},$$

they are said to form a *harmonic* range; P and Q are then *harmonic conjugates* with respect to A and B.

Since

$$\frac{\overrightarrow{PA}}{\overrightarrow{AQ}} = -\frac{\overrightarrow{PB}}{\overrightarrow{BQ}},$$

it follows that *A and B are also harmonic conjugates with respect to P and Q*.

2. Harmonic Pencils

We are to prove a very important theorem, of great generality, proceeding by stages.

(i) THE PROBLEM. Let A, B, P, Q form a harmonic range, so that

$$\frac{\overrightarrow{AP}}{\overrightarrow{PB}} = -\frac{\overrightarrow{AQ}}{\overrightarrow{QB}}.$$

Harmonic Properties

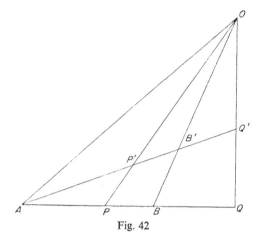

Fig. 42

Take an arbitrary point O and form the pencil of four lines OA, OB, OP, OQ. Let an arbitrary line through A cut the lines of the pencil in A, B', P', Q'. It is required *to prove that A, B', P', Q' also form a harmonic pencil.*

THE DISCUSSION. By the theorem of Menelaus, $PP'O$ is a transversal of $\triangle ABB'$

$$\Rightarrow \frac{\overrightarrow{AP}}{\overrightarrow{PB}} \cdot \frac{\overrightarrow{BO}}{\overrightarrow{OB'}} \cdot \frac{\overrightarrow{B'P'}}{\overrightarrow{P'A}} = -1,$$

and $QQ'O$ is a transversal of $\triangle ABB'$

$$\Rightarrow \frac{\overrightarrow{AQ}}{\overrightarrow{QB}} \cdot \frac{\overrightarrow{BO}}{\overrightarrow{OB'}} \cdot \frac{\overrightarrow{B'Q'}}{\overrightarrow{Q'A}} = -1.$$

Hence

$$\frac{\overrightarrow{AP}}{\overrightarrow{PB}} \cdot \frac{\overrightarrow{B'P'}}{\overrightarrow{P'A}} = -\frac{\overrightarrow{OB'}}{\overrightarrow{BO}} = \frac{\overrightarrow{AQ}}{\overrightarrow{QB}} \cdot \frac{\overrightarrow{B'Q'}}{\overrightarrow{Q'A}}.$$

But

$$\frac{\overrightarrow{AP}}{\overrightarrow{PB}} = -\frac{\overrightarrow{AQ}}{\overrightarrow{QB}},$$

so that

$$\frac{\overrightarrow{AP'}}{\overrightarrow{P'B'}} = -\frac{\overrightarrow{AQ'}}{\overrightarrow{Q'B'}}.$$

(ii) Generalization of the preceding work

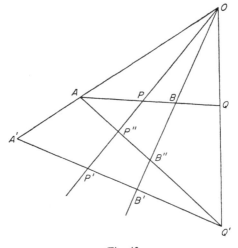

Fig. 43

THE PROBLEM. Once again, let A, B, P, Q, form a harmonic range and take an arbitrary point O. It is required *to prove that, if the pencil OA, OB, OP, OQ is met by* ANY LINE WHATEVER *in points A', B', P', Q', then the range A', B', P', Q' is also harmonic.*

Definition. A pencil of four lines is said to be *harmonic* when it is met by an arbitrary line in the points of a harmonic range.

(By the present theorem, it is sufficient to test the harmonic property for *one* range.)

THE DISCUSSION. Join AQ', meeting OPP' in P'' and OBB' in B''. By theorem (i) just proved,

A, B, P, Q harmonic $\Rightarrow A, P'', B'', Q'$ harmonic,

and argument similar to that used in (i) proves that

A, P'', B'', Q' harmonic $\Rightarrow A', B', P'\ Q'$ harmonic.

Notation: We write

harm. (AB, PQ)

to mean that the range A, B, P, Q is harmonic, with P, Q separating A, B harmonically. Then results such as

harm. (BA, PQ), harm. (PQ, AB), harm. (QP, BA)

follow automatically.

We write similarly

harm. $O(AB, PQ)$

to mean that the pencil OA, OB, OP, OQ is harmonic, with OP, OQ separating OA, OB harmonically.

Problems
1. The internal and external bisectors of $\angle A$ of $\triangle ABC$ meet BC in P, Q. Prove that harm. (BC, PQ).
2. Points D, E, F are taken on BC, CA, AB so that AD, BE, CF are concurrent; $U = EF \cap BC$. Prove that harm. (BC, DU).
3. In $\triangle ABC$, with the notation of Chapter 2, §5, harm. (GH, ON).

3. The Harmonic Tests for Collinearity and Concurrence

(i) THE PROBLEM. Suppose that two harmonic pencils, with vertices U, V have a common "arm", which is necessarily the

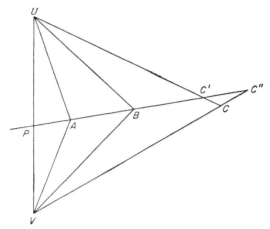

Fig. 44

line UV. It is required *to prove that the three other pairs of corresponding arms intersect in collinear points*; that is, if the intersections are A, B, C, so that

harm. $U(VB, AC)$, harm. $V(UB, AC)$,

then A, B, C are collinear.

THE DISCUSSION. Let $P = AB \cap UV$, $C' = AB \cap UC$, $C'' = AB \cap VC$. Then

harm. $U(PB, AC) \Rightarrow$ harm. (PB, AC'),
harm. $V(PB, AC) \Rightarrow$ harm. (PB, AC'').

Hence C' and C'' coincide, each being the harmonic conjugate of A with respect to P and B. Since their common point must be on UC and on VC, it must be C, so that $C \in AB$.

(ii) THE PROBLEM. Let A, B, P, Q and A, C, X, Y be harmonic ranges on distinct lines, but having a common point A. It is required *to prove that the lines PX, QY, BC are concurrent*.

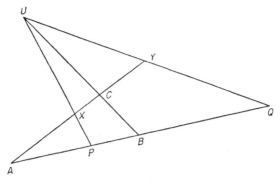

Fig. 45

THE DISCUSSION. Let $U = PX \cap QY$, and suppose that UC meets PQ in a point B' not shown separately in the diagram. Then

harm. $(AC, XY) \Rightarrow$ harm. $U(AC, XY)$
$\Rightarrow$ harm. (AB', PQ).

But we are given that harm. (AB, PQ), so that B' is at B. Hence $U \in BC$.

Problems

1. Two triangles ABC, $A'BC'$ have a common vertex B, the points B, C, C' not being collinear. The bisectors of $\angle A$ meet BC in P, Q; the bisectors of $\angle A'$ meet BC' in P', Q'. Prove that

$$PP' \cap QQ' \in CC', PQ' \cap P'Q \in CC'.$$

2. U is a point in the plane of $\triangle ABC$; $E = BU \cap CA, F = CU \cap AB$; $Y \in BU$ such that harm. (YE, BU) and $Z \in CF$ such that harm. (ZF, CU). Prove that $EF \cap YZ \in BC$.

3. (AB, CD) and $(A'B', C'D')$ are two harmonic ranges on different lines; O, O' are points on the line AA'. Prove that the points $OB \cap O'B'$, $OC \cap O'C'$, $OD \cap O'D'$ are collinear.

4. $ABCD$ is a rectangle; U is the middle point of AD, V is the middle point of BC; $X = UV \cap BD$, $Y = UC \cap BD$. Prove that harm. (BY, XD).

Prove also that, if L is the harmonic conjugate of C with respect to B and V, then $XL \cap YV \in DC$.

5. In $\triangle ABC$, $P \in AB$, $U \in AB$, $Q \in AC$, $V \in AC$ such that harm. (AU, PB) and harm. (AV, QC). Prove that, if $L = BV \cap CU$ and $M = BQ \cap CP$, then $A \in LM$.

Prove also that $PQ \cap UV \in BC$.

4. The Quadrangle

The reader who is new to this work will probably not have given much thought to what he means precisely by the word "quadrilateral". There are, in fact, two similar, but distinct, concepts:

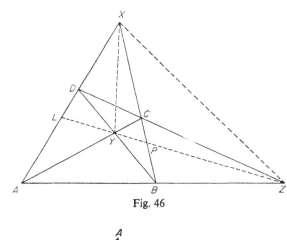

Fig. 46

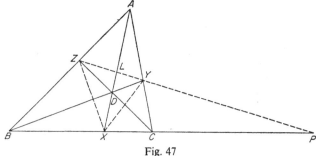

Fig. 47

the quadrangle and the quadrilateral, and it is, indeed, more convenient to begin with the former.

Definitions. The figure defined by four points A, B, C, D and the lines joining them in pairs is called a *quadrangle*, of which A, B, C, D are the *vertices*.

The two diagrams give alternative ways of depicting a typical quadrilateral. They are equally valid for our purposes, but there are sometimes visual advantages in the second, where D is taken inside the triangle ABC.

The six lines

$$BC, AD; \quad CA, BD; \quad AB, CD$$

are called the six *sides* of the quadrangle; we have grouped them in pairs, called *opposite* sides. The points X, Y, Z in which opposite sides meet form a triangle called the *diagonal*, or *harmonic*, triangle of the quadrangle.

The harmonic property of the quadrangle.

The property which follows is basic in the theory of quadrangles and is of very great importance in geometry.

THE PROPERTY. Consider any side, say YZ, of the diagonal triangle of the quadrangle $ABCD$. It meets each of CA, BD at Y and each of AB, CD at Z. Suppose that it meets BC at P and AD at L. It is required *to prove that harm.* (YZ, LP).

THE DISCUSSION. (The second diagram, with D inside the triangle ABC, may be found more convenient for reference.)
PYZ is transversal of $\triangle ABC$

$$\Rightarrow \frac{\overrightarrow{BP}}{\overrightarrow{PC}} \cdot \frac{\overrightarrow{CY}}{\overrightarrow{YA}} \cdot \frac{\overrightarrow{AZ}}{\overrightarrow{ZB}} = -1 \qquad \text{(Theorem of Menelaus)};$$

AX, BY, CZ meet at D

$$\Rightarrow \frac{\overrightarrow{BX}}{\overrightarrow{XC}} \cdot \frac{\overrightarrow{CY}}{\overrightarrow{YA}} \cdot \frac{\overrightarrow{AZ}}{\overrightarrow{ZB}} = +1 \qquad \text{(Theorem of Ceva)}.$$

Hence, comparing these results,

$$\frac{\overrightarrow{BP}}{\overrightarrow{PC}} = -\frac{\overrightarrow{BX}}{\overrightarrow{XC}},$$

so that

$$\text{harm. } (BC, XP).$$

Thus

$$\text{harm. } A(BC, XP)$$
$$\Rightarrow \text{harm. } (ZY, LP)$$
$$\Rightarrow \text{harm. } (YZ, LP).$$

Similar results hold on the sides XY, XZ of the diagonal triangle.

Problems

1. XYZ is the diagonal triangle of the quadrangle $ABCD$; a point D' is taken on AD, and $B' = AB \cap YD'$, $C' = AC \cap ZD'$. Prove that $BC \cap B'C' \in YZ$.

2. XYZ is the diagonal triangle of the quadrangle $ABCD$; $P = BC \cap YZ$, $L = AD \cap YZ$; $U = BL \cap AP$, $V = BY \cap AP$. Prove that $UX \cap VC \in AB$.

3. The altitudes of $\triangle ABC$ are AD, BE, CF, meeting in H. Verify that B, C are two vertices of the diagonal triangle of the quadrangle $AEHF$, and deduce that $EF \cap BC$ is the harmonic conjugate of D with respect to B and C.

4. The incentre and the escribed centres of $\triangle ABC$ are I, I_1, I_2, I_3. Identify the diagonal triangle of the quadrangle I, I_1, I_2, I_3.

5. Duality

There is a large field of work in which the concept known as *duality* is fundamentally important. This book is not greatly

Harmonic Properties

concerned with such aspects, but a brief account must be given here so that the distinction between a quadrangle and a quadrilateral may be clearly understood.

Much of geometry is concerned with the interrelations of points and lines; for convenience, points may be named by capital letters $A, B, C, \ldots$ and lines by lower case letters $a, b, c, \ldots$. Then two points such as A, B define a line, say x; and two lines such as p, q define a point, say U. The essence of the *principle of duality* is that properties of incidence (such as joins and intersections) for a given figure of points $A, B, C, \ldots$ and lines $p, q, r, \ldots$ have an exactly analogous counterpart in a figure of lines $a, b, c, \ldots$ and points $P, Q, R, \ldots$ provided that intersections such as (p, q) are replaced by joins such as PQ while joins such as AB are replaced by intersections such as (a, b).

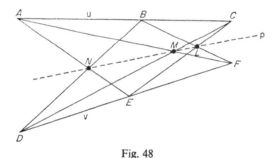

Fig. 48

For example, it can be proved that, *if points A, B, C lie on a line u and points D, E, F on a line v, and if*

$$L = BF \cap CE, M = CD \cap AF, N = AE \cap BD,$$

then L, M, N lie on a line p.

68 *Deductive Geometry*

It is an immediate consequence of the principle of duality, or it can be proved directly, that the theorem obtained by interchanging points and lines, joins and intersections, is also true:

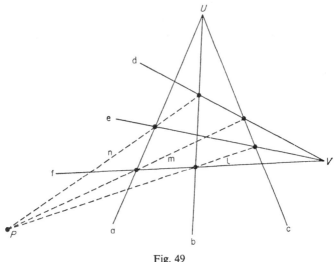

Fig. 49

If lines a, b, c, pass through a point U and lines d, e, f through a point V, and if the line joining the point of intersection (bf) to the point of intersection (ce) is l, with similar notation for m, n, then the lines l, m, n pass through a point P.

6. The Quadrilateral

Definitions. [Compare the dual definitions given in §4.]

The figure formed by four lines *a, b, c, d* and their six intersections when taken in pairs is called a *quadrilateral*, of which *a, b, c, d* are the *sides*.

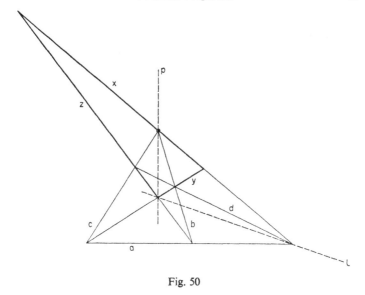

Fig. 50

The six points, which we many conveniently denote by the notation

$$(bc), (ad); (ca), (bd); (ab), (cd)$$

are called the six *vertices* of the quadrilateral; we have grouped them in pairs, called *opposite* vertices. The lines

$$x = \text{join of } (bc), (ad),$$
$$y = \text{join of } (ca), (bd),$$
$$z = \text{join of } (ab), (cd)$$

form a triangle known as the *diagonal* or *harmonic* triangle of the quadrilateral.

Quadrangle and quadrilateral together.

The accompanying diagram can be interpreted to give both a quadrangle and a quadrilateral. As a *quadrangle*, it has vertices

A, B, C, D and sides BC, AD, CA, BD, AB, CD, with XYZ as diagonal triangle; as a *quadrilateral*, it has AB, BC, CD, DA as sides, A, B, C, D, X, Z as vertices, and UYW as diagonal triangle.

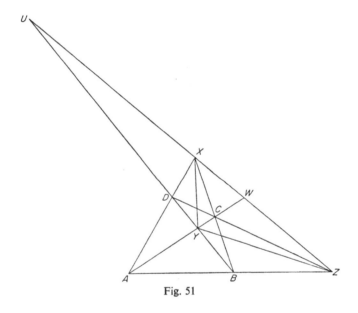

Fig. 51

Thus the two diagonal triangles are different, but they have a common vertex Y and side $XZUW$.

The harmonic property of the quadrilateral

THE PROPERTY. The vertex (yz) of the diagonal triangle (Fig. 50) lies on the line joining (ca), (bd) and on the line joining (ab), (cd). Let its join to (bc) be the line p and its join to (ad) be the line l. It is required *to prove that harm.* (yz, lp).

THE DISCUSSION. This is actually identical with the property already proved for the quadrangle; for, in the notation of Fig. 51,

Harmonic Properties

what we have to prove is that harm. $Y(XZ, UW)$, and this is an immediate consequence of the quadrangle property for the side ZX of the diagonal triangle and its intersections with the lines CA, BD.

COROLLARY. In the "joint" diagram of Fig. 51, the vertices X, Z of the diagonal triangle of the quadrangle are separated harmonically by the vertices U, W of the diagonal triangle of the quadrilateral.

7. Some Illustrations
(i) The polar line of a point with respect to a triangle

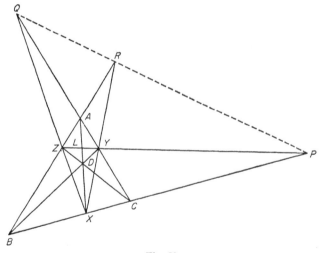

Fig. 52

THE PROPERTY. Let ABC be a given triangle and D a point in its plane. Let $X = BC \cap AD, Y = CA \cap BD, Z = AB \cap CD$; and then let $P = BC \cap YZ, Q = CA \cap ZX, R = AB \cap XY$. It is required *to prove that P, Q, R are collinear.*

72   *Deductive Geometry*

Definition. The line PQR is called the *polar line* of D with respect to the triangle ABC.

THE DISCUSSION. If $L = AD \cap YZ$, then XYZ is diagonal triangle of quadrangle $ABCD$ ⇒ harm. (LP, YZ) ⇒ harm. $X(LP, YZ)$ ⇒ harm. (AC, YQ).
Similarly harm. (AB, ZR).
But
$$\left.\begin{array}{l}\text{harm. } (AC, YQ) \\ \text{harm. } (AB, ZR)\end{array}\right\} \Rightarrow CB, YZ, QR \text{ collinear}$$

Thus CB, YZ meet on QR, that is,
$$P \in QR.$$

(ii) The angle bisectors of a triangle

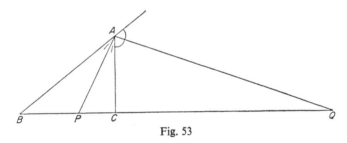

Fig. 53

THE PROBLEM. Given a triangle ABC, let the internal and external bisectors of the angle A meet BC at P, Q. It is required *to prove that harm.* (BC, PQ).

THE DISCUSSION:

$$AP \text{ is internal bisector of } \angle A \Rightarrow \frac{\overrightarrow{BP}}{\overrightarrow{PC}} = +\frac{AB}{AC};$$

Harmonic Properties

AQ is external bisector of $\angle A \Rightarrow \dfrac{\overrightarrow{BQ}}{\overrightarrow{QC}} = -\dfrac{AB}{AC}.$

Hence
$$\dfrac{\overrightarrow{BP}}{\overrightarrow{PC}} = -\dfrac{\overrightarrow{BQ}}{\overrightarrow{QC}},$$
so that harm. (BC, PQ).

8. The Bisection Theorems for a Harmonic Pencil

(i) THE PROBLEM. The pencil $U(AC, BD)$ is given to be harmonic, with B, D separating A, C.

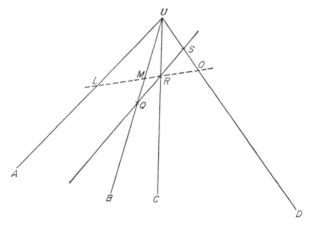

Fig. 54

An arbitrary line is drawn parallel to UA, cutting UB, UC, UD in Q, R, S. It is required *to prove that R is the middle point of QS.*

THE DISCUSSION. Draw any line through R cutting UA, UB, UD in L, M, O. Then

74   *Deductive Geometry*

$$\text{harm. } U(AC, BD) \Rightarrow \text{harm. } (LR, MO)$$
$$\Rightarrow \frac{\overrightarrow{LM}}{\overrightarrow{MR}} = -\frac{\overrightarrow{LO}}{\overrightarrow{OR}},$$

so that, numerically,

$$\frac{LM}{MR} = \frac{LO}{OR}.$$

But

$$QR \| LU \Rightarrow \frac{LM}{MR} = \frac{LU}{QR}$$

and

$$RS \| LU \Rightarrow \frac{LO}{OR} = \frac{LU}{RS}.$$

Hence

$$\frac{LU}{QR} = \frac{LU}{RS},$$

so that

$$QR = RS.$$

(ii) THE PROBLEM. [The converse of Illustration (ii) of §7.]

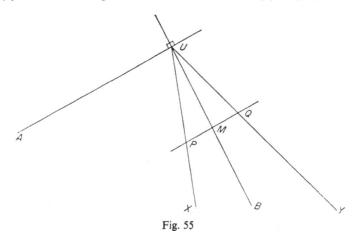

Fig. 55

Harmonic Properties

Let $U(AB, XY)$ be a harmonic pencil, with UX, UY separating UA, UB, and such that
$$AU \perp UB.$$
It is required *to prove that UA, UB are the bisectors of angle XUY.*

Let a line parallel to UA meet UX, UB, UY in P, M, Q. Then, by the preceding,
$$\left. \begin{array}{r} \text{harm. } U(AB, XY) \\ PMQ \| AU \end{array} \right\} \Rightarrow PM = MQ.$$

Also
$$\left. \begin{array}{r} PMQ \| AU \\ UM \perp AU \end{array} \right\} \Rightarrow UM \perp PQ.$$

In triangles UMP, UMQ:
$$UM = UM, MP = MQ, \angle UMP = \angle UMQ,$$
so that
$$\triangle UMP \equiv \triangle UMQ,$$
so that
$$\angle PUM = \angle QUM.$$

Hence UM is a bisector of $\angle PUQ$, and so AU, being perpendicular to UM, is the other bisector.

Theorems

1. The converse of §8(i) is also true:
   If $U(ABCD)$ is a given pencil, and if a line parallel to UA cuts UB, UC, UD in Q, R, S so that $QR = RS$, then harm. (AC, BD).

Problems

1. Given a parallelogram $ABCD$ and a line AU such that $AU \| BD$, prove that harm. $A(BD, UC)$.
2. $ABCD$ is a parallelogram; P, Q are the middle points of AB, AD. Prove that harm. $B(AD, QC)$ and harm. $D(AB, PC)$, and deduce that $BQ \cap DP \in AC$.
3. XYZ is the diagonal triangle of the quadrangle $ABCD$. The line

through *Y* parallel to *XZ* cuts *BC*, *AB*, *AD*, *CD* in *L*, *N*, *P*, *R*. Prove that *LN* = *PR*.

4. Points *D*, *E*, *F* are taken on the sides *BC*, *CA*, *AB* of △*ABC* so that *AD*, *BE*, *CF* meet in a point *O*; *U* = *EF* ∩ *BC*. The line through *D* parallel to *AU* cuts *AB*, *AC* in *M*, *N*; the line through *D* parallel to *OU* cuts *OB*, *OC* in *QR*. Prove that *MQNR* is a parallelogram.

5. The diagonals *AC*, *BD* of a parallelogram meet in *O*; through *O* are drawn lines *l*, *m* so that *l*∥*AB*, *m*∥*AD*, and any point *R* is taken on *m*. The line *RC* meets *l* in *P* and *DB* in *U*; the line *RD* meets *l* in *Q* and *AC* in *V*. Prove that harm. (*RP*, *CU*) and harm. (*RQ*, *DV*), and deduce that *UV*∥*AB*.

9. A Test for a Harmonic Range

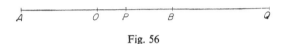

Fig. 56

THE PROBLEM. Let *A*, *P*, *B*, *Q* be four collinear points, and let *O* be the middle point of *AB*. It is required *to prove that*

$$\text{harm. } (AB, PQ) \Leftrightarrow OA^2 = OP \cdot OQ,$$

the arrow of consequence going in both senses.

THE DISCUSSION. Since *O* is the middle point of *AB*,

$$\overrightarrow{AO} = \overrightarrow{OB}.$$

Now

$$\text{harm. } (AB, PQ) \Leftrightarrow \frac{\overrightarrow{AP}}{\overrightarrow{PB}} = -\frac{\overrightarrow{AQ}}{\overrightarrow{QB}}$$

$$\Leftrightarrow \overrightarrow{AP} \cdot \overrightarrow{QB} + \overrightarrow{AQ} \cdot \overrightarrow{PB} = 0$$

$$\Leftrightarrow (\overrightarrow{AO} + \overrightarrow{OP})(\overrightarrow{QO} + \overrightarrow{OB})$$

$$+ (\overrightarrow{AO} + \overrightarrow{OQ})(\overrightarrow{PO} + \overrightarrow{OB}) = 0$$

$$\Leftrightarrow (\overrightarrow{OB} + \overrightarrow{OP})(\overrightarrow{OB} - \overrightarrow{OQ})$$
$$+ (\overrightarrow{OB} + \overrightarrow{OQ})(\overrightarrow{OB} - \overrightarrow{OP}) = 0$$
$$\Leftrightarrow \overrightarrow{OB}^2 + \overrightarrow{OB}(\overrightarrow{OP} - \overrightarrow{OQ}) - \overrightarrow{OP}\cdot\overrightarrow{OQ}$$
$$+ \overrightarrow{OB}^2 - \overrightarrow{OB}(\overrightarrow{OP} - \overrightarrow{OQ}) - \overrightarrow{OP}\cdot\overrightarrow{OQ} = 0$$
$$\Leftrightarrow 2\overrightarrow{OB}^2 - 2\overrightarrow{OP}\cdot\overrightarrow{OQ} = 0$$
$$\Leftrightarrow \overrightarrow{OA}^2 = \overrightarrow{OB}^2 = \overrightarrow{OP}\cdot\overrightarrow{OQ}.$$

Problems

1. A, B are two given points; a circle is drawn through A, B and a point P is taken on it; the tangent at P meets the line AB in U. Prove that the circle of centre U and radius UP cuts AB in points X, Y such that harm. (AB, XY).

2. In $\triangle ABC$, $AD \perp BC$, $BE \perp CA$, $CF \perp AB$, $P = EF \cap BC$ and A' is the middle point of BC. Prove that $A'E$ is the tangent at E to $\odot EDP$.

3. In $\triangle ABC$, $\angle A = 90°$ and $AD \perp BC$. The circle of centre B and radius BA cuts BC in U, V. Prove that harm. (UV, BC).

4. In $\triangle ABC$, $\angle A = 90°$, $AD \perp BC$ and the tangent at A to $\odot ABC$ meets BC in X. Prove that harm. (BC, DX).

5. A, B, P, Q are four collinear points such that harm. (AB, PQ). An arbitrary circle through P, Q cuts the circle on AB as diameter in U, V. Prove that OU, OV are the tangents from O to $\odot PUQV$.

6. l is a given line and A, B two points not on it, both on the same side of l. Prove that $\exists$ two points U, $V \in l$ such that l is a tangent to $\odot ABU$ and to $\odot ABV$.

An arbitrary circle through A, B cuts l in P, Q. Prove that harm. (PQ, UV).

7. Two circles cut orthogonally. A diameter AB of one cuts the other in points P, Q. Prove that harm. (AB, PQ).

SIX
Pole and Polar

1. The Polar of a Point with Respect to a Circle

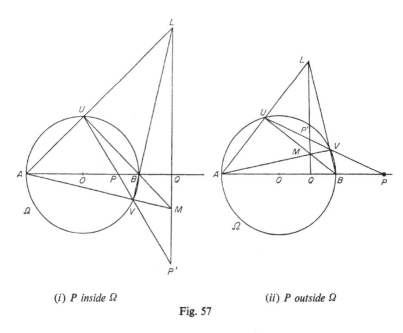

(i) *P* inside Ω (ii) *P* outside Ω

Fig. 57

Definitions. Two points P, P' in the plane of a circle Ω of centre O are said to be *conjugate* with respect to Ω if the line PP' meets the circle in points U, V such that harm. (UV, PP').

Pole and Polar

[We shall have to extend this definition later to cover the cases where the line PP' does not meet the circle.]

When P is fixed, the locus of points P' such that P, P' are conjugate with respect to Ω is called the *polar* of P with respect to Ω. The point P itself is called the *pole* of its polar.

THE PROBLEM. It is required *to prove that the polar of P is a straight line perpendicular to the diameter through P.*

THE DISCUSSION. Draw the diameter through P, meeting Ω in the points A, B. Let an arbitrary chord through P meet the circle in U, V and let P' be the harmonic conjugate of P with respect to U, V; then P' lies on the polar of P. Finally, let Q be the harmonic conjugate of P with respect to A, B; in particular, Q lies on the polar of P.

Now

$\left.\begin{array}{l}\text{harm. } (UV, P'P) \\ \text{harm. } (AB, QP)\end{array}\right\} \Rightarrow UA, VB, P'Q$ concurrent, say in L.

Also

$\left.\begin{array}{l}\text{harm. } (UV, P'P) \\ \text{harm. } (BA, QP)\end{array}\right\} \Rightarrow UB, VA, P'Q$ concurrent, say in M.

Further,

AB diameter of Ω
$\Rightarrow BU \perp AL, AV \perp BL$
$\Rightarrow AV, BU$ meet in the orthocentre of $\triangle ABL$
$\Rightarrow M$ is the orthocentre of $\triangle ABL$
$\Rightarrow LM \perp AB$
$\Rightarrow P'Q \perp AB$.

Thus P' lies on the line through the *fixed* point Q perpendicular to the *fixed* line AB, so that the locus of P' is † the straight line through Q perpendicular to AB.

† But see §2.

Note:

$$\left.\begin{array}{l}\text{harm. } (AB, PQ) \\ O \text{ middle point of } AB\end{array}\right\} \Rightarrow OA^2 = OP \cdot OQ$$

$\Rightarrow P, Q$ inverse points with respect to Ω.

Hence *the polar of P is the line through the inverse of P with respect to Ω and perpendicular to the diameter through P.*

COROLLARY. It is an immediate consequence of the definition that *if the polar of a point P passes through a point P', then the polar of P' passes through P*. For if the line PP' cuts† the circle Ω in points U, V, then harm. $(UV, PP') \Leftrightarrow$ harm. $(UV, P'P)$.

2. The Case when P is Outside the Circle Ω

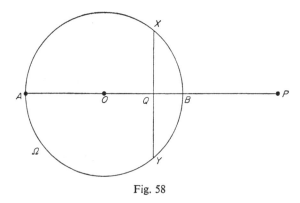

Fig. 58

In §1 we ignored a complication, with which we now deal, occurring when P lies outside the circle Ω. But first we prove a theorem of some importance in itself.

† But see §2.

Pole and Polar

THE PROBLEM. Let the polar of a point P outside the circle Ω meet the circle in points X, Y. It is required *to prove that PX, PY are the tangents from P to Ω.*

THE DISCUSSION. As in §1, P, Q are inverse points with respect to Ω. Hence
$$OP \cdot OQ = OX^2$$
$$\Rightarrow \frac{OX}{OP} = \frac{OQ}{OX}$$
so that, since also $\angle XOP = \angle QOX$,
$$\triangle XOP \sim \triangle QOX$$
$$\Rightarrow \angle OXP = \angle OQX$$
$$= 90°$$
$$\Rightarrow XP \perp OX$$
$$\Rightarrow XP \text{ is the tangent at } X.$$

Similarly YP is the tangent at Y.

The difficulty in the stated definition for the polar of P is that an arbitrary line through P may not meet the circle at all: indeed, it will not do so unless it lies "within" the angle XPY. The locus of P', as defined, is only *that part of the straight line which lies between X and Y*. We have nevertheless adopted the given definition since it is hoped that the reader will later come to study complex projective geometry where that definition stands and where the difficulty does not arise (because of so-called "imaginary" points of intersection of line and circle).

To resolve the dilemma, such as it is, we can replace the given definition by the equivalent property that P' is on the line through the inverse of P and perpendicular to the diameter through P. All positions of P' are then accounted for, even when the line PP' does not meet Ω.

We must check that, in the excluded case, it is still true that, *if the polar of P passes through P', then the polar of P' passes through P.* (The proof in §1 depended on the existence of *U, V*.)

Suppose that P' is on the polar of P. Then $OP'\ OQ = a^2$, and $P'Q \perp OP'$, where Q is the inverse of P with respect to Ω. Now let Q' be the inverse of P'. Then

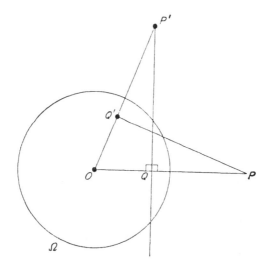

Fig. 59

$OP \cdot OQ = a^2 = OP' \cdot OQ'$
$\Rightarrow PQ\ Q'P'$ cyclic
$\Rightarrow PQ' \perp OP'$
$\Rightarrow P$ is on the polar of P'.

The result is therefore true whether or not the line PP' meets the circle Ω.

Theorem

1. A, B, C, D are four points on a circle, and XYZ is the polar triangle of the quadrangle $ABCD$. Then $\triangle XYZ$ is such that each side is the polar of the opposite vertex. (Compare §3 below.)

Problems

1. A, B, P are three collinear points. Prove that the polars of P with respect to all circles through A, B have a common point.
2. Prove that the angle between the polars of A and B is equal to the angle subtended at the centre by AB.
3. A, B, C, D are four collinear points such that harm. (AB, CD); their polars with respect to a circle Ω are a, b, c, d. Prove that the lines a, b, c, d form a pencil such that harm. (ab, cd).
4. The sides of $\triangle ABC$ are the polars of the vertices of $\triangle PQR$ with respect to a circle Ω. Prove that the sides of $\triangle PQR$ are the polars of the vertices of $\triangle ABC$.
5. Given two points A, B and a circle Ω of centre O; draw $BX \perp$ polar of A and $AY \perp$ polar of B. Prove that $AY/AO = BX/BO$.

[A possible method is to draw $OM \perp AY$ and $OL \perp BX$; then $\triangle AMO \sim \triangle BLO$.]

Corollary. If $AY = 0$, so that A is on the polar of B, then $BX = 0$, so that B is on the polar of A.

3. Self-Polar (Self-Conjugate) Triangle with Respect to Ω

THE PROBLEM. Let P be any point (inside or outside the circle) in the plane of a circle Ω of centre O, and let Q be any point (inside or outside the circle) on the polar of P. Let the polars of P, Q meet in R. It is required *to prove that PQ is the polar of R.*

THE DISCUSSION. The proof is direct:

R is on the polar of P ⇒ the polar of R is through P;

R is on the polar of Q ⇒ the polar of R is through Q.

Hence the polar of R is PQ.

Definition. A triangle such as PQR, in which each side is the polar of the opposite vertex, is said to be *self-polar* or *self-*

conjugate with respect to Ω. Each two vertices are then conjugate with respect to Ω. The following result is immediate:

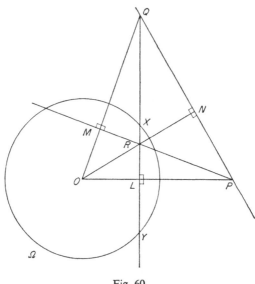

Fig. 60

THE PROBLEM. *To prove that the four points O, P, Q, R are so related that each is the orthocentre of the triangle formed by the other three.*

THE DISCUSSION. QR is the polar of $P \Rightarrow OP$ is perpendicular to QR. Similarly OQ, OR are perpendicular to RP, PQ. By definition of orthocentre, this is the required result.

We now prove a basic property of self-polar triangles:

THE PROBLEM. It is required *to prove that a triangle self-polar with respect to a circle is necessarily obtuse-angled.*

Pole and Polar

THE DISCUSSION. Observe first that a triangle self-polar with respect to Ω must have at least two vertices outside Ω; for if there is one vertex at all, say R, inside Ω, then the polar of R lies entirely outside Ω, so that both P and Q must be outside.

If, say, P is one of the outside vertices, then its polar passes through the points of contact X, Y of the tangents from P to Ω and, since Q, R lie on XY and are harmonically conjugate with respect to X, Y, one of them, say Q, is outside and the other, R, is therefore inside. Let OP, OQ, OR meet QR, RP, PQ in L, M, N; then L, P; M, Q; N, R are inverse pairs with respect to Ω, so that L, P are on the same side of O; M, Q are on the same side of O; N, R are on the same side of O. When, however, $\triangle PQR$ is acute, O lies between L, P, between M, Q, and between N, R, so this case is not possible. Hence $\triangle PQR$ is obtuse-angled.

4. Harmonic Pencils on a Circle

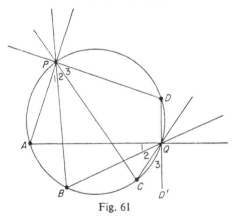

Fig. 61

(i) We start with a lemma:

THE PROBLEM. Let A, B, C, D be four fixed points and P a variable point on a given circle. It is required *to prove that the* pencil *PA, PB, PC, PD remains constant, in the sense that the angles between those four lines (produced in both directions) have the same values for all positions of P.*

THE DISCUSSION. Suppose, for example, that *P* is on the arc *AD* as shown. The angles *APB, BPC, CPD* remain constant for all such positions of *P*.

If, however, *P* moves into, say, the arc *CD*, to the position marked *Q*, then, if *DQ* is produced beyond *Q* to *D'*,

$\angle AQB = \angle APB$ (same segment)
$\angle BQC = \angle BPC$ (same segment)
$\angle CQD' = \angle CPD$ (external angle theorem)

so that the pencils can be superposed, in the way implied by the diagram. Corresponding angles are therefore equal.

Note: The result is also true when *P* is at *A, B, C* or *D*, provided that (for, say, *P* at *A*) the line *PA* is interpreted as the tangent at *A*.

The proof is the same as in the general case save that, if *AT* is the tangent at *A*,

$\angle TAB = \angle APB$

because of the theorem on the angle between tangent and chord.

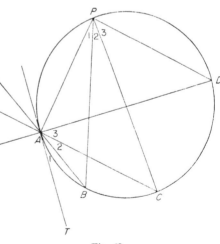

Fig. 62

(ii) Harmonic separation on a circle

We have seen what is meant by harmonic separation on a straight line, and this concept must now be extended from straight line to circle.

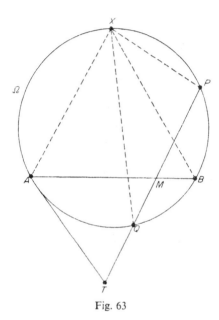

Fig. 63

THE PROBLEM. Let A,B, P,Q be four given points on a circle Ω. If X is a varying point of the circle, then we have just proved that the pencil $X(AB, PQ)$ remains constant in shape for all positions of X. What we are to prove is that *the pencil $X(AB, PQ)$ is harmonic when the four points are so related that each of the chords AB, PQ passes through the pole of the other.*

THE DISCUSSION. Having in mind the note at the end of (i), let the tangent at A meet PQ in T. Let $M = PQ \cap AB$. Then

pencil harmonic $\Rightarrow$ harm. $A(TB, PQ)$
$\Rightarrow$ harm. (TM, PQ).

Moreover, if the tangent at B meets PQ in T', then, in the same way,

harm. $(T'M, PQ)$,

so that T, T' are the same point.

Thus PQ passes through T, the pole of AB; and, consequently, AB also passes through the pole of PQ.

Note: The converse theorem is also true:

If AB, PQ are chords such that each passes through the pole of the other, then $X(AB, PQ)$ is a harmonic pencil for any position of X on Ω.

It is, indeed, sufficient to take X at A, so that we may refer again to Fig. 63, where T is now known to the pole of AB. Then

T is pole of $AB \Rightarrow$ harm. (TM, PQ)
$\Rightarrow$ harm. $A(TM, PQ)$
$\Rightarrow$ harm. $A(AB, PQ)$
$\Rightarrow$ harm. $X(AB, PQ)$

Definitions. Two chords such that each passes through the pole of the other are said to be *conjugate* with respect to Ω.

Four points such as A, B, P, Q which subtend a harmonic pencil at every point of the circle (so that, as just proved, the chords AB, PQ are conjugate) are said to be *harmonic* on the circle; we also say that A, B *separate* P, Q *harmonically*.

Theorem

1. XYZ is the diagonal triangle of a cyclic quadrilateral $ABCD$ (notation of p. 64). The tangents at B, C meet in U, the tangents A, D meet in V, the tangents at A, B meet in L, the tangents at C, D meet in M. Then Y, Z, U, V are collinear and X, Y, L, M are collinear.

Problems

1. AB is a diameter of a circle and CD a chord perpendicular to it. Prove that harm. (AB, CD) on the circle.

Deduce that, if $U = BC \cap AD$, $V = AC \cap BD$, then the tangents at C and D meet on UV.

2. The tangents at points A, B on a circle meet in T and a line through T meets the circle in UV. Prove that, if $L = AU \cap TB$ and $M = AV \cap TB$, then harm. (TB, LM).

Prove also that, if $X = BU \cap TA$, and $Y = BV \cap TA$, then $LX \cap MY \in AB$ and $LY \cap MX \in AB$.

3. The chords UV, PQ of a circle meet in T and A, B are the points of contact of the tangents from T to the circle. Prove that $PU \cap QV \in AB$ and $PV \cap QU \in AB$.

4. A point U is taken on a circle of which AB is a diameter. The tangents at B, U meet in T, and AU meets BT in S. By first proving that the lines AU, AB divide harmonically the line AT and the tangent at A, prove that T is the middle point of BS.

Prove this result also by elementary geometry.

5. AB is a diameter of a circle of centre O. A line through B cuts this circle in U and also cuts the circle Ω on OB as diameter in V. The line through O parallel to UB cuts Ω again in L; the line UO cuts the circle Ω again in M. Prove that harm. (LV, MB) on Ω.

SEVEN
Line and Plane

1. Preliminary Ideas

A detailed study of the properties of lines and planes in space is lengthy and, at this stage, somewhat tedious. We propose to pass lightly over some of the things that intuition, not necessarily correctly, regards as obvious, reserving closer study for those matters that are more likely to be found troublesome.

The basic concepts are the *point* and the *straight line*, which we assume to be familiar. From them we derive the *plane*, which is defined to be a surface such that, if A, B are any two points whatever upon it, then the line AB lies wholly in it. Whether such a surface can exist effectively is not as clear as might be thought, but that is a consideration over which we do not linger.

We enunciate without proof a number of *propositions of incidence*, whose truth is to be regarded as evident. Certain complications arising from ideas of parallelism are ignored for the present. (But see below, p. 91.)

(i) Two lines lying in a plane have a point in common.

(ii) Two lines meeting in a point lie in a plane; all the points of that plane can be constructed by taking two variable points, one on each given line, and drawing the line joining them (see (iii) below).

(iii) A unique line joins any two points in space.

(iv) Two planes in space intersect in the points of a straight line.

(v) A straight line and a plane meet, in general, in a single point.

Line and Plane

(vi) Given a point and a line, there is, in general, a unique plane passing through each of them.

(vii) Three planes meet, in general, in a single point.

(viii) Three non-collinear points determine a unique plane.

Notation: We shall often use capital letters A, B, C, ... to denote points, small italic letters a, b, c, ... to denote lines, and Greek letters α, β, γ, δ, ... (alpha, beta, gamma, delta, ...) to denote planes.

Thus two planes α, β might meet in a line u which, in its turn, might meet a plane γ in a point P.

2. Parallel Lines and Skew Lines

Let l, m be two given straight lines. According to the properties listed in §1, they lie in a plane if they meet and they cannot lie in a plane if they do not meet.

We did, however, indicate in §1 that those properties, while perfectly true in general, were subject to certain exceptions. These we must now study.

Suppose that l, m do indeed lie in a plane. They necessarily intersect, with the one case of exception: they may be *parallel*. We are therefore led to consider two distinct types of non-intersecting lines:

Definitions. Two straight lines which lie in a plane but do not meet are called *parallel*.

Two straight lines which do not lie in a plane (cannot meet and) are called *skew*.

For example, in the "box" shown in the diagram,

$$AD\|BC\|A'D'\|B'C',$$
$$AB\|DC\|A'B'\|D'C',$$
$$AA'\|BB'\|CC'\|DD'.$$

Examples of pairs of skew lines are:

$$AB, CC'; A'B', DD'; CD, BB'.$$

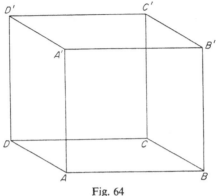

Fig. 64

Theorems

1. If u, v are two skew lines and p, q are two lines each meeting each of them, then p, q are skew.
2. Given two skew lines u, v and a point O not on either, a unique line (a *transversal* of u, v) can be drawn through O to meet both u and v.
3. Given three mutually skew lines u, v, w, a line can be drawn through an arbitrary point of u to cut v and w (a *transversal* of u, v, w). Any two such lines are skew.
4. Two lines each parallel to a third are parallel to one another.
5. If two triangles ABC, $A'B'C'$ in different planes are so related that AA', BB', CC' have a common point O (the triangles then being said to be *in perspective*), the points $L = BC \cap B'C'$, $M = CA \cap C'A'$, $N = AB \cap A'B'$ exist and L, M, N are collinear. (Theorem of *Desargues*.)

Problems

1. A, B, C, D, O are five points in general position in space. A transversal from O meets DB in M and CA in Q; another meets DC in N and AB in R. Prove that $MN \cap QR \in BC$, $NQ \cap MR \in AD$.
2. Three mutually skew lines u, v, w are each met by each of three other mutually skew lines p, q, r. Notation such as $p \cap u$ denotes the point common to p, u; notation such as $p \wedge u$ denotes the plane containing p, u. By considering the lines of intersection of the planes $p \wedge u$,

$q \wedge v$, $r \wedge w$ taken two at a time, prove that the three lines $(q \cap w, r \cap v)$, $(r \cap u, p \cap w)$, $(p \cap v, q \cap u)$ are concurrent.

3. If u, v are two skew lines and a line p parallel to u meets v while a line q parallel to v meets u, prove that p, q are skew.

4. A line l meets a plane π in a single point O. Prove that l is skew to every line in π not passing through O.

3. The Angle Between Two Lines

Let AB, CD be two skew lines. The angle between them is defined as follows:

Definition. Let O be an arbitrary point, and draw the lines through O parallel to AB, CD. *The angle between AB and CD is defined to be the angle between these two (coplanar) lines through O.*

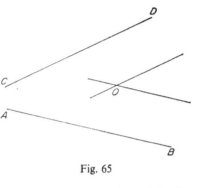

Fig. 65

THE PROBLEM. It is required *to prove that the angle so defined is independent of the position of O.*

THE DISCUSSION. If O' is an alternative position, choose a point P and a point Q on the two lines through O; on the corresponding parallel lines through O', choose P', Q' so that $OP' = OP$, $OQ' = OQ$. Then $OO'P'P$, $OO'Q'Q$ are parallelograms, so that PP', QQ' (being equal and parallel to OO') are equal and parallel;

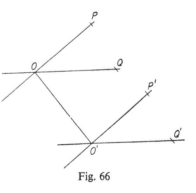

Fig. 66

hence $PQQ'P'$ is a parallelogram, so that $PQ = P'Q'$.

Hence
$$\triangle OPQ \equiv \triangle O'P'Q' \quad (3 \text{ sides})$$
$$\Rightarrow \angle POQ = \angle P'OQ'.$$

Note: There are *two* angles between the lines through O, being supplementary angles. In practice, this seldom causes confusion.

4. Perpendicular Lines and Normals to Planes

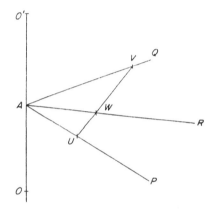

Fig. 67

(i) THE PROBLEM. Let OA be a given line, and AP, AQ two lines each perpendicular to it. The lines AP, AQ define a plane and it is required *to prove that OA is perpendicular to* every *line through A in the plane.*

THE DISCUSSION. Let AR be any line through A in the plane APQ. We have to prove that $OA \perp AR$.

Produce OA its own length to O' and let an arbitrary line be drawn to cut AP in U and AQ in V. Then:

UV meets AP, AQ $\Rightarrow$ UV is in the plane APQ
$\Rightarrow$ UV meets AR, say in W.

Then
$$\triangle OAU \equiv \triangle O'AU \Rightarrow OU = O'U \brace \triangle OAV \equiv \triangle O'AV \Rightarrow OV = O'V} \Rightarrow \triangle OUV \equiv \triangle O'UV.$$

But
$$\triangle OUV \equiv \triangle O'UV \Rightarrow \angle OUW = \angle O'UW,$$

and
$$\left. \begin{array}{r} \angle OUW = \angle O'UW \\ OU = O'U \\ UW = UW \end{array} \right\} \Rightarrow \triangle OUW \equiv \triangle O'UW$$
$$\Rightarrow OW = O'W$$
$$\Rightarrow OA \perp AW \text{ (since } OA = O'A)$$

Note that the *converse result* is also true:—

(ii) THE PROBLEM. A line OA is given in space, and AP, AQ, AR are three distinct lines perpendicular to it. It is required *to prove that AP, AQ, AR are coplanar*.

THE DISCUSSION. The lines AP, AQ define one plane and the lines OA, AR define another (different from the first since OA cannot have two distinct lines perpendicular to it and lying with it in one plane). Let the two planes meet in a line AS. Then

$$\left. \begin{array}{l} OA \perp AP, OA \perp AQ \\ AS \text{ in plane } PAQ \end{array} \right\} \Rightarrow OA \perp AS.$$

But
$$\left. \begin{array}{l} OA, AR, AS \text{ coplanar} \\ OA \perp AR, OA \perp AS \end{array} \right\} \Rightarrow AR, AS \text{ coincide}.$$

Hence AR is in the plane APQ.

Definition. Given a point A in a plane, a line through A is said to be *perpendicular* or *normal* to the plane if it is perpendicular to every line through A in the plane.

Test: The line will, in fact, be perpendicular to the plane if it is perpendicular to two *lines through A in the plane.*

This follows from the preceding work.

(iii) Existence of a normal

THE PROBLEM. Given a point A in a plane, it is necessary *to establish the existence of a line which is normal at A to the plane.*

THE DISCUSSION. Let AP, AQ be two lines through A in the plane. Draw the plane through A having AP as normal (by drawing two lines through A perpendicular to AP) and the plane through A having AQ as normal. Let these planes meet in a line AO.

Then

AO in plane with AP as normal $\Rightarrow AO \perp AP$

AO in plane with AQ as normal $\Rightarrow AO \perp AQ$

and

$$\left. \begin{array}{l} OA \perp AP \\ OA \perp AQ \end{array} \right\} \Rightarrow OA \perp \text{plane } APQ,$$

so that OA is the required normal.

Theorem

1. There cannot be two distinct lines through a point A normal to a given plane, whether or not A is in the plane.

Problems

1. In the figure of the cube (p. 92) find the angles between the following pairs of lines:

   (i) AB, CC'; (ii) AC, $A'B'$; (iii) BD, $A'C'$; (iv) $A'D$, $B'C'$.

2. Prove that an arbitrary point P on the normal to a plane π at a point A in the plane is equidistant from every point of any given circle of centre A lying in π.

3. Three points A, B, C in a plane π are equidistant from a point O not in π. The centre of $\odot ABC$ is U. Prove that $OU \perp \pi$.

5. Parallel Planes

Definitions. Two planes which do not meet, however far they are produced in any direction, are said to be *parallel*.

Remark: Strictly speaking, we do not assert that such planes can exist; we have merely provided a name for them if they do.

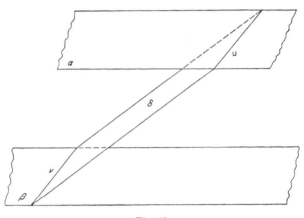

Fig. 68

(i) THE PROBLEM. Let α, β be two given parallel planes and let δ be an arbitrary plane meeting them in lines u, v. It is required *to prove that $u \| v$*.

THE DISCUSSION. The lines u, v cannot meet, otherwise each plane α, β would pass through their common point so that the planes could not be parallel. Also, u, v are coplanar, lying in δ. Hence $u \| v$.

(ii) THE PROBLEM. It is required *to prove that two planes which have a common point O have a whole line in common*.

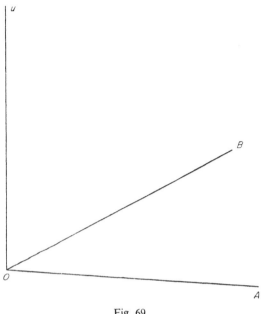

Fig. 69

THE DISCUSSION. Let the given planes α, β (not shown in the diagram) have a common point O, and let OA, OB be the normals at O to α, β. Let u be the line perpendicular to OA and OB.

Then

$$u \perp OA \Rightarrow u \in \alpha,$$
$$u \perp OB \Rightarrow u \in \beta.$$

Hence α, β meet in the line u.

(iii) THE PROBLEM. The actual existence of parallel planes can be confirmed by giving a precise construction:

Let α be a given plane and A a point not in it. It is required *to define a plane through A parallel to α.*

Line and Plane

Take two non-parallel lines u, $v \in \alpha$ and write $P = u \cap v$. In the plane containing A, u draw the line m through A parallel

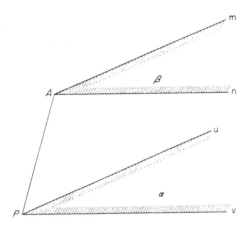

Fig. 70

to u; in the plane containing A, v draw the line n through A parallel to v. Then *the plane β through m, n is parallel to α*.

THE DISCUSSION. If the planes α, β are not parallel, they meet, by (ii), in a line x. Now the plane through m, u meets x in a point M which, being by definition in α, must lie on u and which, being by definition in β, must lie on m. But $u \| m$, so that M cannot exist. Thus x, u lie in α but do not meet; hence $x \| u$. Similarly $x \| v$. But u, v are not parallel, and so the existence of x is contradicted.

Hence

$$\beta \| \alpha.$$

Definition. A line u is said to be *parallel* to a plane α if there is no point common to u and α.

Theorem

1. If $u \| \alpha$, then an arbitrary plane through u meets α in a line parallel to u.

(iv) The following result enables us *to draw a straight line u through a given point A parallel to a given plane* δ. (There are many solutions, all such lines u lying in the plane through A parallel to δ. See the Theorem below.)

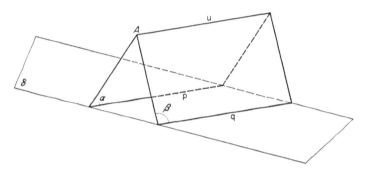

Fig. 71

THE PROBLEM. Let p, q be two parallel lines lying in the plane δ. Draw the planes through A, p and A, q, meeting in a line u, necessarily through A. It is required *to prove that the line $u \|$ the plane δ*.

THE DISCUSSION. Suppose that u does meet δ, say in a point O. Then O, being on u, is also in each of the planes α, β. But

$$\left. \begin{array}{l} O \text{ in } \alpha \\ O \text{ in } \delta \end{array} \right\} \Rightarrow O \text{ on } p,$$

$$\left. \begin{array}{l} O \text{ in } \beta \\ O \text{ in } \delta \end{array} \right\} \Rightarrow O \text{ on } q.$$

Line and Plane

But O cannot be on each of the parallel lines p, q, so that the assumption that O exists is false. Thus u does not meet δ, so that $u \| \delta$.

Theorem

1. The lines through a given point A parallel to a plane α line in a plane β parallel to α.

6. Properties of Normals

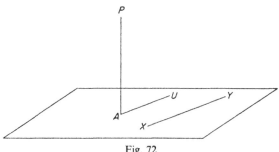

Fig. 72

(i) THE PROBLEM. Let AP be the normal to a given plane at a point A. We have proved that AP is perpendicular to every line *through A* in the plane. We now observe that, more generally, *AP is perpendicular to* EVERY *line in the plane.*

THE DISCUSSION. Let XY be a line in the plane, not through A, and draw AU, in the plane, parallel to XY.

Then, by definition (§3), the angle between AP and XY is equal to the angle between AP and AU. Since AP is normal to the plane, this is a right angle.

(ii) THE PROBLEM. In a similar way we prove that *a line AP is normal to a given plane if it is perpendicular to* ANY TWO *(nonparallel) lines in the plane.*

THE DISCUSSION. Draw through the point *A*, where the given line meets the plane, the two lines parallel to the lines in the plane. Then, as in (i), *AP* is perpendicular to each of these lines and therefore normal to the plane.

(iii) THE PROBLEM. Two *perpendicular* skew lines *AB*, *PQ* are given. It is required *to prove that there exists a unique plane through PQ having AB as normal*.

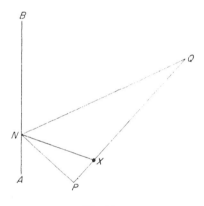

Fig. 73

THE DISCUSSION. Let *X* be an arbitrary point of *PQ*, and draw $XN \perp AB$. Then

$$\left.\begin{array}{r}AB \perp PQ\\ AB \perp NX\end{array}\right\} \Rightarrow AB \perp \text{plane } NPQ.$$

The required plane is thus determined.

Also it is unique; for if there were a second plane, cutting *AB* in *N'*, then $XN' \perp AB$, which is impossible unless $N' = N$.

(iv) The theorem of the three perpendiculars

THE PROBLEM. Let *AB* be normal at *A* to a given plane and *PQ* an arbitrary line in the plane. It is required *to prove that, if $BM \perp PQ$, then $AM \perp PQ$ or, alternatively, that, if $AM \perp PQ$, then $BM \perp PQ$.*

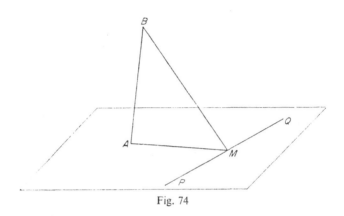

Fig. 74

THE DISCUSSION.

AB normal to the plane $\Rightarrow AB \perp PQ \Rightarrow PQ \perp AB$.

Then

$$\left.\begin{array}{l} PQ \perp AB \\ PQ \perp BM \end{array}\right\} \Rightarrow PQ \perp \text{plane } ABM$$
$$\Rightarrow PQ \perp AM.$$

Conversely,

$$\left.\begin{array}{l} PQ \perp AB \\ PQ \perp AM \end{array}\right\} \Rightarrow PQ \perp \text{plane } ABM$$
$$\Rightarrow PQ \perp BM.$$

(v) THE PROBLEM. It is required *to prove that all normals to a given plane are parallel.*

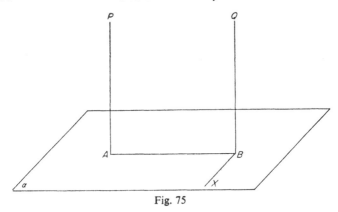

Fig. 75

THE DISCUSSION. Let AP, BQ be normal at A, B to a given plane α. Draw BX in the plane α so that $XB \perp AB$.

Then, by the theorem of the three perpendiculars,

$$\left.\begin{array}{r} PA \perp \alpha \\ AB \perp XB \end{array}\right\} \Rightarrow PB \perp XB.$$

Thus XB is perpendicular to BA, BP, BQ, so that those three lines are coplanar. In particular, AP, BQ are coplanar and, each being perpendicular to AB, are therefore parallel.

6. The Common Perpendicular of Two Skew Lines

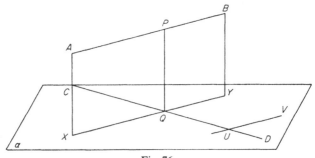

Fig. 76

Line and Plane

THE PROBLEM. *Two skew lines AB, CD are given. It is required to construct a line PQ meeting them (with $P \in AB$ and $Q \in CD$) so that PQ is perpendicular to both AB and CD.*

THE DISCUSSION. Draw through CD the plane α parallel to AB; this is done by drawing through any point U of CD the line UV parallel to AB and taking the plane $CDUV$. Draw $AX, BY \perp \alpha$, and let XY meeet CD in Q. Then *the line through Q perpendicular to AB is the required common perpendicular, meeting AB in P.*

The proof is immediate:

$$AB \parallel \alpha \Rightarrow AB \parallel XY$$
$$\left.\begin{array}{l} AB \parallel XY \\ AX \perp XY, BY \perp XY \end{array}\right\} \Rightarrow ABYX \text{ is a rectangle}$$
$$\Rightarrow XA \perp AB.$$

Also $ABYX$ is a rectangle and PQ is in its plane, so that

$$\begin{array}{l} XA \perp AB \\ QP \perp AB \end{array} \Rightarrow PQ \parallel AX \Rightarrow PQ \perp \alpha$$
$$\Rightarrow PQ \perp CD.$$

Theorems

1. If A, B are two given points, the locus of a point P such that $PA = PB$ is a plane to which AB is perpendicular.
2. If A, B, C are three given points, the locus of a point P such that $PA = PB = PC$ is, in general, a line perpendicular to the plane ABC.
3. There is, in general, one single point equidistant from four given points A, B, C, D.
4. (*A converse of the theorem of the three perpendiculars*). Given a point B and a plane α not through it, if PQ is a line in α and $BM \perp PQ$, and if, further, MX is drawn in α so that $XM \perp PQ$, then the line BA such that $BA \perp XM$ is the perpendicular from B to α.
5. If a line u is normal to two distinct planes α, β, then $\alpha \parallel \beta$.
6. The common perpendicular of two skew lines is the shortest distance between them.
7. Two straight lines with two distinct common perpendiculars are parallel.

Problems

1. The common perpendicular of two skew lines u, v meets u in A and v in B. Prove that the locus in space of points equidistant from A, B is a plane α such that $u\|\alpha$, $v\|\alpha$.

Prove that $P \in u$, $Q \in v \Rightarrow$ the middle point of $PQ \in \alpha$.

2. Given three mutually skew lines u, v, w, prove that a line can be found meeting them in U, V, W such that U is the middle point of VW.

3. Given two skew lines u, v and variable points $P \in u$, $Q \in v$, prove that the locus of the middle point of PQ is a plane to which the common perpendicular of u, v is normal.

4. The common perpendicular of two perpendicular skew lines u, v meets u in P and v in Q. Points A, B are taken on u so that P is the middle point of AB. Prove that every point of v is equidistant from A and B.

5. Given four skew lines, u, v, u', v' such that $u\|u'$, $v\|v'$, prove that the common perpendicular of u, v is parallel to the common perpendicular of u', v'.

6. Three parallel planes α, β, γ are given; u is a line in α, v a line in β, w a line in γ. Prove that, if β lies between α and γ, then the sum of the common perpendiculars of u, v and of v, w is equal to the common perpendicular of u, w.

EIGHT
Some Standard Solid Bodies

1. The Parallelepiped

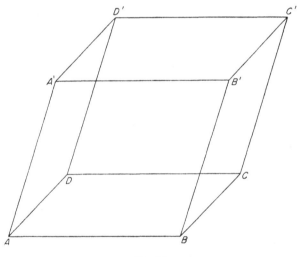

Fig. 77

(The purpose of this account is to make the reader familiar with the basic properties of the figures. The definitions are presented in a correct logical order, but proofs are sketched rather than given in detail since that seems more appropriate to the present level of work.)

(i) Three pairs of parallel planes serve to define a "skew box" known as a parallelepiped. In the diagram, the pairs of parallel planes are

$$ABCD \| A'B'C'D',$$
$$ADD'A' \| BCC'B',$$
$$ABB'A' \| DCC'D'.$$

These six planes determine the *faces* of the parallelepiped, in each of which is a parallelogram cut out by the four faces not parallel to it.

There are eight *vertices* A, B, C, D, A', B', C', D' and twelve *edges* grouped in three parallel sets:

$$AB \| DC \| A'B' \| D'C',$$
$$AD \| BC \| A'D' \| B'C',$$
$$AA' \| BB' \| CC' \| DD'.$$

There are four *diagonals* AC', BD', CA', DB', and they *bisect each other* at a point O known as the *centre* of the parallelepiped.

(ii) THE RECTANGULAR PARALLELEPIPED (or "box"). If the three angles at A are all right angles, then all the angles of all the faces are right angles and the parallelepiped is called *rectangular*.

A special feature of this case is that *the four diagonals are all equal*. A sphere can be drawn with centre O to pass through the eight vertices.

(iii) THE CUBE. If the three edges through A of a rectangular parallelepiped are all equal, then all the twelve edges are equal, and the figure is called a *cube*.

2. The Tetrahedron

Four planes in general position serve to define a figure known

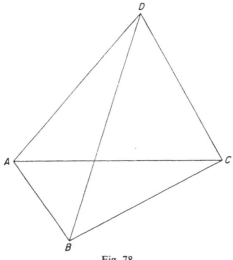

Fig. 78

as a *tetrahedron*. It has four *faces*, each cut in a triangle by the other three, so that the faces are

$$DBC,\ DCA,\ DAB,\ ABC.$$

There are six *edges*, grouped in three *opposite* pairs:

$$AD,\ BC;\ BD,\ CA;\ CD,\ AB.$$

The four points A, B, C, D are called *vertices*.

Problems

1. A parallelepiped is named as in Fig. 79 so that A, A'; B, B'; C, C'; D, D' are pairs of opposite vertices. Verify that the parallelepiped can be split up into the five tetrahedra

$$A'BCD + AB'CD + ABC'D + ABCD' + ABCD.$$

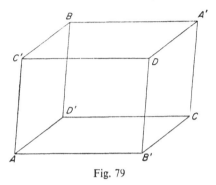

Fig. 79

2. Prove, in the diagram, that $AD'A'D$ is a parallelogram. Hence prove that AA', BB', CC', DD' have a common point O.

3. Prove that, if the parallelepiped is rectangular, then

$$AA'^2 = AB'^2 + AC'^2 + AD'^2.$$

4. Prove that the middle points of AB', $B'C$, CD', $D'A$, $A'B$, BC', $C'D$, DA' are at the vertices of a parallelepiped.

3. Centroid Properties of the Tetrahedron

Consideration of the middle points of the edges of a tetrahedron leads to an exciting chain of properties. Denote by P, Q, R the middle points of BC, CA, AB and by L, M, N the middle points of AD, BD, CD.

(i) THE PROPERTIES. We are *to prove that*

(a) *QRMN, RPNL, PQLM are parallelograms, whose sides are parallel to the appropriate edges of the tetrahedron.*

(b) *LP, MQ, NR bisect each other at a point G.*

Some Standard Solid Bodies

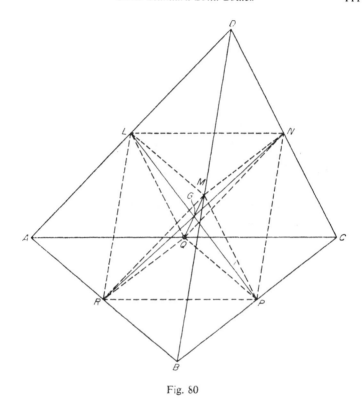

Fig. 80

THE DISCUSSION.

$$\left.\begin{array}{l}AR = RB \\ AQ = QC\end{array}\right\} \Rightarrow QR \| BC \text{ and } QR = \tfrac{1}{2} BC$$

$$\left.\begin{array}{l}DM = MB \\ DN = NC\end{array}\right\} \Rightarrow MN \| BC \text{ and } MN = \tfrac{1}{2} BC.$$

Hence

$$QR \| MN, \; QR = MN$$
$$\Rightarrow QRMN \text{ is a parallelogram.}$$

Similarly

RPNL, PQLM are parallelograms

Further

QRMN is a parallelogram

⇒ *QM, RN* have a common middle point *G*.

Also

RPNL is a parallelogram

⇒ *LP, RN* have a common middle point, which is also *G*.

Thus *LP, MQ, NR* bisect each other at *G*.

(ii) An alternative treatment links these properties from a somewhat different point of view

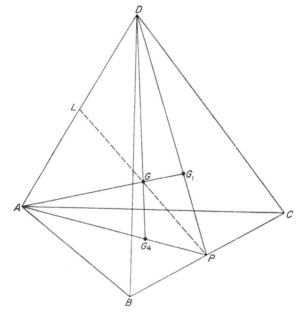

Fig. 81

Some Standard Solid Bodies 113

THE PROBLEM. Let P be the middle point of BC. Then the centroids of $\triangle DBC$, $\triangle ABC$ are points G_1, G_4 on DP, AP such that $PG_4 = \frac{1}{3} PA$, $PG_1 = \frac{1}{3} PD$. It is required *to prove that, if points G_2, G_3 are defined similarly for $\triangle CAD$, $\triangle ABD$, then AG_1, BG_2, CG_3, DG_4 have a common point G.*

THE DISCUSSION. Let AG_1 meet DG_4 in a point which we shall call G.

Then

$$\left. \begin{array}{l} PG_4 = \frac{1}{3} PA \\ PG_1 = \frac{1}{3} PD \end{array} \right\} \Rightarrow G_4 G_1 \| AD \text{ and } G_4 G_1 = \frac{1}{3} AD.$$
$$\Rightarrow G_4 G = \frac{1}{3} GD \text{ and } G_1 G = \frac{1}{3} GA.$$

Hence AG_1 meets DG_4 in the points of quadrisection furthest from A and D.

Similarly BG_2, CG_3, defined in the same way, pass likewise through G.

Thus AG_1, BG_2, CG_3, DG_4 *have a common point G which is, for each, the point of quadrisection farthest from the corresponding vertex of the tetrahedron.*

(iii) THE PROBLEM. We have *to identify the point G as defined in (ii) with the point G as defined in (i)*. In case (ii), let $L = PG \cap AD$. Then it can be proved by ratios, or by the theorem of Ceva, that L is the middle point of AD; for example,

$$\frac{\overrightarrow{AG_4}}{\overrightarrow{G_4 P}} \cdot \frac{\overrightarrow{PG_1}}{\overrightarrow{G_1 D}} \cdot \frac{\overrightarrow{DL}}{\overrightarrow{LA}} = 1$$

$$\Rightarrow \frac{3}{1} \cdot \frac{1}{3} \cdot \frac{\overrightarrow{DL}}{\overrightarrow{LA}} = 1$$

$$\Rightarrow DL = LA.$$

114 *Deductive Geometry*

Thus G, as defined in (ii), lies on AP and, similarly, on BQ, CR. It is thus the same as G as defined in (i).

Definition. The point G is called the *centroid* of the tetrahedron.

4. Orthocentral Properties of the Tetrahedron

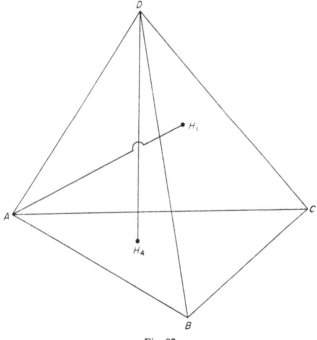

Fig. 82

Definition. The lines AH_1, BH_2 CH_3, DH_4 drawn from the vertices of a tetrahedron perpendicular to the opposite faces are called the *altitudes* of the tetrahedron.

Some Standard Solid Bodies 115

The properties of the altitudes with which we shall be concerned presuppose a specialisation of the tetrahedron which must now be considered. The tetrahedron will be of the type known as *orthogonal*.

THE PROBLEM. Suppose that the tetrahedron has the property that two pairs of opposite edges are perpendicular: say $BD \perp CA$, $CD \perp AB$. It is required *to prove that the remaining edges are also perpendicular, so that $AD \perp BC$*.

Definition. A tetrahedron whose opposite edges are perpendicular is called *orthogonal*.

THE DISCUSSION. Referring to Fig. 80,

$$BD \perp CA \Rightarrow LR \perp LN$$
$$\Rightarrow LNPR \text{ is a rectangle}$$
$$\Rightarrow LP = NR.$$
$$CD \perp AD \Rightarrow MP \perp ML$$
$$\Rightarrow MLQP \text{ is a rectangle}$$
$$\Rightarrow LP = MQ.$$

Hence
$$MQ = NR$$
$$\Rightarrow NMRQ \text{ is a rectangle}$$
$$\Rightarrow NQ \perp NM$$
$$\Rightarrow AD \perp BC.$$

For the rest of this paragraph, we assume that the tetrahedron ABCD is orthogonal.

(i) THE PROBLEM. Let *ABCD* be an orthogonal tetrahedron, so that

$$AD \perp BC, BD \perp CA, CD \perp AB.$$

It is required *to prove that the four altitudes* AH_1, BH_2, CH_3, DH_4 *meet in a point H.*

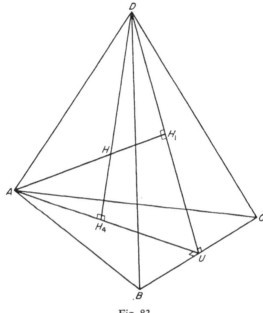

Fig. 83

Definition. The point H is called the **orthocentre** of the tetrahedron.

THE DISCUSSION. Draw $AU \perp BC$. Then

$$\left.\begin{array}{r}BC \perp AU \\ BC \perp AD\end{array}\right\} \Rightarrow BC \perp \text{plane } ADU \Rightarrow BC \perp DU$$

Thus $DU \perp BC$.

Now draw $AH_1 \perp DU$, $DH_4 \perp AU$, and let AH_1, DH_4, in the plane ADU, meet in H. Then

$$BC \perp \text{plane } ADU \Rightarrow BC \perp DH_4,$$

and
$$\left.\begin{array}{l}DH_4 \perp BC\\ DH_4 \perp AU\end{array}\right\} \Rightarrow DH_4 \perp ABC.$$

Thus DH_4 is the altitude from D; similarly AH_1 is the altitude from A.

In other words, AH_1 meets DH_4.

Similarly BH_2 meets DH_4. If, then, BH_2 does *not* pass through H, it must lie in the plane of AH_1 and DH_4: that is, in the plane ADU, which is impossible. Hence BH_2 passes through H. Similarly CH_3 passes through H, so that AH_1, BH_2, CH_3, DH_4 all pass through H.

COROLLARY. *H_1, H_2, H_3, H_4 are the orthocentres of* $\triangle DBC$, $\triangle DCA$, $\triangle DAB$, $\triangle ABC$. For H_4 is on the altitude from A, and similar argument would have obtained it on the altitudes from B, C; and similarly for the other triangles.

(ii) There is a test for an orthogonal tetrahedron in terms of the lengths of the sides

THE PROBLEM. It is required *to prove that*

ABCD is orthogonal $\Leftrightarrow DA^2 + BC^2 = DB^2 + CA^2 = DC^2 + AB^2.$

THE DISCUSSION. By (i),
$ABCD$ orthogonal $\Rightarrow AU \perp BC,\ DU \perp BC$
$\qquad\qquad\qquad\quad \Rightarrow AB^2 - AC^2 = BU^2 - UC^2 = DB^2 - DC^2$
$\qquad\qquad\qquad\quad \Rightarrow DB^2 + CA^2 = CD^2 + AB^2,$

and the result follows.

If, conversely, $DB^2 + CA^2 = CD^2 + AB^2,$
then
$$DB^2 - DC^2 = AB^2 - AC^2,$$
so that, if
$$DU_1 \perp BC,\ AU_2 \perp BC,$$

it follows that
$$U_1 \equiv U_2 \equiv U, \text{ say.}$$

Thus
$$BC \perp DU, BC \perp AU$$
$$\Rightarrow BC \perp \text{plane } ADU$$
$$\Rightarrow BC \perp AD.$$

Similarly
$$CA \perp BD, AB \perp CD.$$

Theorems

1. If O_1, O_2, O_3, O_4 are the circumcentres of $\triangle DBC$, $\triangle DCA$, $\triangle DAB$, $\triangle ABC$, then the lines through O_1, O_2, O_3, O_4 perpendicular to the planes containing those points have a common point O which is equidistant from A, B, C, D.

Definition. The point O is called the *circumcentre* of the tetrahedron $ABCD$.

2. The diagram shows the section of an orthogonal tetrahedron $ABCD$ by the plane through D and the Euler line $O_4 G_4 H_4$ of $\triangle ABC$. The circumcentre and centroid of the tetrahedron are O, G and OG meets DH_4 in a point temporarily called H^*. Prove that $GH^* = OG$ and deduce (by symmetry of argument) that H^* is the orthocentre H of the tetrahedron.

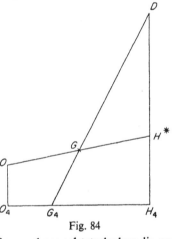
Fig. 84

3. The middle point of the edges of an orthogonal tetrahedron lie on a sphere.

Problems

1. (Notation of §3). Prove that, if $AD = BC$, $BD = CA$, $CD = AB$, then LP, MQ, NR are mutually orthogonal.

By proving first that $\triangle LBC$ and $\triangle PAD$ are both isosceles, or otherwise, prove that LP is the common perpendicular of AD, BC.

2. Given a tetrahedron $ABCD$ and a point O, the transversal from O to BC, AD cuts BC in P and AD in L; the transversal from O to CA, BD cuts CA in Q and BD in M; the transversal from O to AB, CD cuts AB in R and CD in N. Prove that AP, BQ, CR meet on DO, and that BN, CM, DP meet on AO.

3. Given a tetrahedron $ABCD$, points M, N are taken on CA, AB and points Q, R are taken on DB, DC so that $MN \| BC$, $QR \| BC$. Prove that $NQ \cap MR \in AD$.

4. A plane meets the edges of a tetrahedron in six points. Prove that they are the vertices of a quadrilateral in the plane.

The tetrahedron is $ABCD$ and the plane meets AB in P, BC in Q, CD in R, DA in S. Prove that (having regard to sign)

$$\frac{AP}{PB} \cdot \frac{BQ}{QC} \cdot \frac{CR}{RD} \cdot \frac{DS}{SA} = 1.$$

State and prove the converse of this property.

5. $ABCD$ is a tetrahedron in which $AB = AC$, $DB = DC$. Prove that $AD \perp BC$.

Prove also that any point of AD is equidistant from B and C.

6. $ABCD$ is a tetrahedron in which $DA = DB = DC$. Prove that the foot of the perpendicular from O to the plane ABC is the centre of $\odot ABC$.

NINE
Angles between Lines and Planes.

WE HAVE already (p. 93) considered the angle between two skew lines. We pass now to some further definitions.

1. The Angle Between a Line and a Plane
Let u be a given line and α a given plane meeting it in a point A.

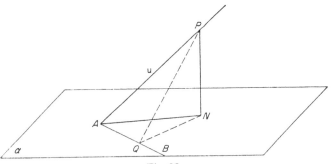

Fig. 85

Definition. The angle between the line u and the plane α is defined to be the angle NAP, where N is the foot of the perpendicular to the plane from an arbitrary point P of the line.

It is an elementary exercise in similar triangles to verify that this angle is independent of the position of P on u.

THE PROBLEM. Let AB be an arbitrary line through A in the plane α. It is required *to prove that*
$$\angle PAN < \angle PAB$$

and that

$$\angle BAN < \angle PAB.$$

THE DISCUSSION. Draw $PQ \perp AB$. By the theorem of the three perpendiculars, $NQ \perp AB$.

Thus

$PN < PQ$ (right-angled triangle NPQ)
$$\Rightarrow \angle PAN < \angle PAQ$$

and

$QN < PQ$ (right-angled triangle NPQ)
$$\Rightarrow \angle QAN < \angle PAQ.$$

2. The Angle Between Two Planes

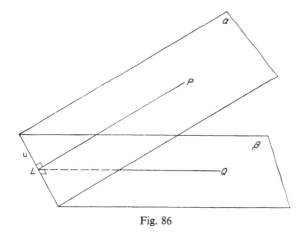

Fig. 86

Let α, β be two given planes meeting in a straight line u.

Definition. The angle between the two planes is defined to be the angle PLQ, where L is an arbitrary point of u and LP, LQ are the lines in α, β perpendicular to u.

122 *Deductive Geometry*

This angle is independent of the point L selected, since, if $P'L'Q'$ is an alternative position,

$$\left.\begin{array}{l} LP, L'P' \perp u \Rightarrow L'P' \| LP \\ LQ, L'Q' \perp u \Rightarrow L'Q' \| LQ \end{array}\right\} \Rightarrow \angle P'L'Q' = \angle PLQ.$$

3. The 'Line of Greatest Slope' Property

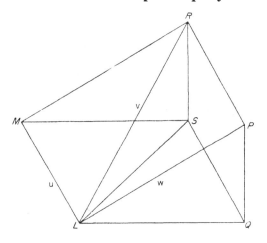

Fig. 87

THE PROBLEM. For convenience of reference, regard β in §2 as the horizontal plane, and consider a line v in the plane α. It is required *to prove that the angle between the plane α and the horizontal is greater than the angle between v and the horizontal*: in other words, if v meets the common line u of α, β in L, and if LP in α is perpendicular to u, then the line LP makes a greater angle with the horizontal than v.

THE DISCUSSION. Let P be an arbitrary point of the line w through L perpendicular to u; draw $PQ \perp \beta$ and let $PR \| u$ cut v in R; draw $RS \perp \beta$.

Then $PRSQ$ is a rectangle, so that $RS = PQ$, and also, since $\angle LPR = 90°$, $LR > LP$.

Thus, in the language of trigonometry,

$$\frac{PQ}{LP} > \frac{RS}{LR}$$
$$\Rightarrow \sin \angle PLQ > \sin \angle RLS$$
$$\Rightarrow \angle PLQ > \angle RLS.$$

Definition. The lines in the plane α which are perpendicular to u are called the *lines of greatest slope* of the plane.

Remark: The language of trigonometry is not essential to the argument, but it helps to make statements more concise.

Problems

1. $ABCD$ is a regular tetrahedron. Prove that, if θ is the angle between DA and the plane ABC, then $\cos \theta = 1/\sqrt{3}$ and that, if ϕ is the angle between the planes DBC, ABC, then $\cos \phi = \frac{1}{3}$.

2. A cube has two parallel square faces $ABCD$, $A'B'C'D'$, so that the edges perpendicular to those faces are AA', BB', CC', DD'. The face $ABCD$ is horizontal. Prove that
   (i) the plane $DAB'C'$ makes an angle of 45° with the horizontal,
   (ii) if AC' makes an angle θ with the horizontal, then $\sin \theta = 1/\sqrt{3}$,
   (iii) if the plane $D'AC$ makes an angle ϕ with the horizontal, then $\cos \phi = 1/\sqrt{3}$.

3. $ABCD$ is a tetrahedron in which $BC = CA = AB = 4$, $DB = DC = 5$, $DA = 3$. Prove that, if θ is the angle between the planes DBC, ABC, then $\cos \theta = 2/\sqrt{7}$, and that, if ϕ is the angle between DB and the plane ABC, then $\cos \phi = \frac{4}{5}$.

4. $ABCD$ is a tetrahedron in which $\angle BDC = \angle CDA = \angle DAB = 90°$, and $DA = a$, $DB = b$, $DC = c$. Prove that the angle between AD and the plane ABC is θ, where $\tan \theta = bc/a\sqrt{(b^2 + c^2)}$, and prove that the angle between the planes DBC, ABC is $90° - \theta$.

Verify that the tetrahedron is orthogonal.

5. $ABCD$ is a tetrahedron in which $BC = CA = AB$ and in which also $DA = DB = DC$. Prove that, if the line AD makes an angle of 60° with the plane ABC, then the angle between the planes DBC, ABC is ϕ, where $\tan \phi = 2\sqrt{3}$.

4. Angles at a Point

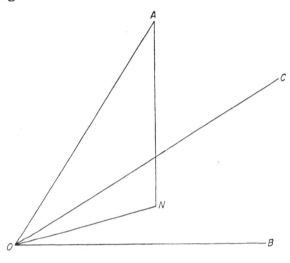

Fig. 88

Let three planes meet at a point O, their lines of intersection in pairs being OA, OB, OC. The three plane angles BOC, COA, AOB form a unit known as a *trihedral angle*.

Two properties are important:

(i) THE PROBLEM. It is required *to prove that the sum of any two of $\angle BOC$, $\angle COA$, $\angle AOB$ is greater than the third*: say

$$\angle AOB + \angle AOC > \angle BOC.$$

THE DISCUSSION. Draw $AN \perp$ plane BOC. Then (pp. 120–1)

$$\angle AOB > \angle NOB,$$
$$\angle AOC > \angle NOC,$$

so that

$$\angle AOB + \angle AOC > \angle NOB + \angle NOC.$$

Angles Between Lines and Planes

If, as in the diagram (Fig. 88), ON lies in the angle BOC, it follows at once that

$$\angle NOB + \angle NOC = \angle BOC,$$

so that

$$\angle AOB + \angle AOC > \angle BOC.$$

If ON lies outside the angle BOC, then one or other of $\angle NOB$, $\angle NOC$ is greater than $\angle BOC$, so that

$$\angle NOB + \angle NOC > \angle BOC,$$

and, again,

$$\angle AOB + \angle AOC > \angle BOC.$$

Note: The case of *equality* occurs only when OA, OB, OC are coplanar with OA "inside" the angle BOC. Then

$$\angle AOB + \angle AOC = \angle BOC.$$

(ii) THE PROBLEM. It is required *to prove that*

$$\angle BOC + \angle COA + \angle AOB < 360°.$$

By what we have just proved,

$$\angle OAB + \angle OAC > \angle BAC,$$
$$\angle OBC + \angle OBA > \angle CBA,$$
$$\angle OCA + \angle OCB > \angle ACB.$$

Add and rearrange:

$$(\angle OBC + \angle OCB) + (\angle OCA + \angle OAC) + $$
$$+ (\angle OAB + \angle OBA) > \angle BAC + \angle CBA + \angle ACB,$$

so that

$$(180° - \angle BOC) + (180° - \angle COA) + (180° - \angle AOB)$$
$$> 180°,$$

or

$$\angle BOC + \angle COA + \angle AOB < 360°.$$

5. The Five Regular (Platonic) Solids

THE PROBLEM. The regular tetrahedron has all its faces equilateral triangles and the cube has all its faces squares. It is a matter of interest *to prove that there are only five convex solids whose faces are regular polygons*.

THE DISCUSSION. The possible regular polygons, with the sizes of corresponding angles, are, in the first instance,

triangle	$60°$
square	$90°$
pentagon	$108°$
hexagon	$120°$, etc.

the angles increasing with the number of sides.

Now there must be at least 3 faces at a vertex, and there may be more. On the other hand, the theorem of §4 can be extended to prove that the sum of the angles at a vertex is in all cases less than $360°$. Hence, if there are n faces meeting at a typical vertex, we have

$$n \geqslant 3$$

and

triangles, $60n < 360,$
squares, $90n < 360,$
pentagons, $108n < 360,$
hexagons, $120n < 360.$

The possibilities are thus:

triangles	3, 4, 5 at a vertex
squares	3 at a vertex
pentagons	3 at a vertex.

The figures defined in this way are called the *regular* or *Platonic* solids. Several books may be consulted for further details. In particular, the reader will find an excellent account of how to

Angles Between Lines and Planes 127

construct them in *Mathematical Models* by H. Martyn Cundy and A. P. Rollett, Clarendon Press, 1952.

The regular tetrahedron (triangles, $n = 3$) and the cube (squares, $n = 3$) are familiar. The following diagrams of the other solids are based, with permission which we gratefully acknowledge, on *Mathematical Models*.

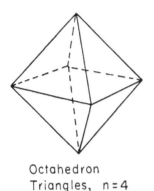

Octahedron
Triangles, $n = 4$

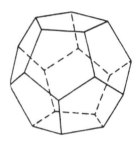

Dodecahedron
Pentagon, $n = 3$

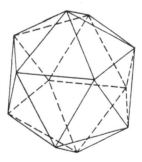

Icosahedron
Triangles, $n = 6$

Fig. 89

Problems

1. Copy the cube shown in the diagram and mark the middle points of the edges.

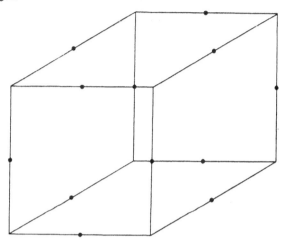

Fig. 90

A new solid is formed by removing the 8 tetrahedra each of which has as its four vertices one vertex of the cube and the three middle points nearest to it. Verify that the resulting solid has 12 vertices, 24 edges, 14 faces of which 6 are square and 8 triangular. Prove also that the sum of the angles at any vertex is 300°.

2. Copy the regular tetrahedron shown in the diagram and mark the points of trisection of the edges.

A new solid is formed by removing the 4 tetrahedra each of which has as its

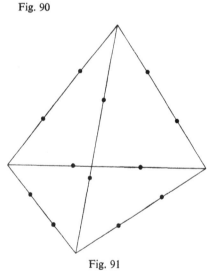

Fig. 91

four vertices one vertex of the given tetrahedron and the three points of trisection nearest to it. Verify that the resulting solid has 12 vertices, 18 edges, 8 faces of which 4 are hexagonal and 4 triangular. Prove that the sum of the angles at any vertex is 300°.

Remark: A famous theorem due to Euler states that, for any such convex body, the number of vertices + the number of faces exceeds by 2 the number of edges.

TEN
The Sphere

1. Definition and First Properties

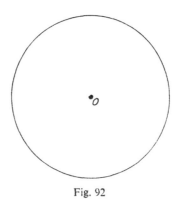

Fig. 92

Definitions. The *sphere* is a surface traced in space by a point whose distance from a fixed point O (the *centre*) has a constant value (the *radius*). Any chord through the centre is called a *diameter* and any plane through the centre is called a *diametral plane*.

(i) THE PROBLEM A general plane may or may not meet the sphere. It does meet it when its distance from the centre is less

The Sphere 131

than the radius. It is required *to prove that a plane meeting the sphere does so in the points of a circle.*

THE DISCUSSION. Let α be a given plane in the presence of a sphere of centre O and radius a. Draw $OA \perp \alpha$, and let $OA = p$. Then the plane cuts the sphere if $p < a$.

Now let P be any point common to α and the sphere.

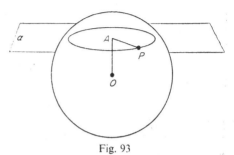

Fig. 93

Then $OA \perp \alpha \Rightarrow OA \perp AP$
$$\Rightarrow AP^2 = OP^2 - OA^2 = a^2 - p^2 = \text{constant.}$$

Thus P, lying in α, is at constant distance $\sqrt{(a^2 - p^2)}$ from the fixed point A, so that the locus of P is a circle.

(ii) The tangent plane at a point Q

As a particular case of (i), suppose that $p = a$. Then the foot of the perpendicular from O to the plane is on the sphere, at a point which we now call Q. The plane α is the *tangent plane at Q*, meeting the circle at the point Q, and only at Q. The radius OQ is perpendicular to every line in the tangent plane at Q.

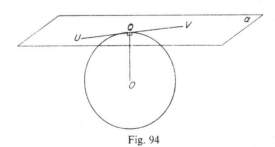

Fig. 94

Theorems

1. Two circles in different planes but having two points in common define a sphere on which both lie.
2. If two spheres pass through a point A and have the same tangent plane there, the distance between their centres is either the sum or the difference of their radii.

Definition. The two spheres are said to *touch* at A.

3. Two spheres which intersect do so in the points of a circle; the distance between their centres is less than the sum of their radii.

Problems

1. The line joining the centres of two circles cut on a sphere by parallel planes is perpendicular to each plane.
2. The centres of the circles of given radius on a sphere lie on a concentric sphere.
3. The centres of the circles on a sphere whose planes pass through a fixed point A lie on the sphere having OA as a diameter, where O is the centre of the given sphere.
4. The larger the radius of a circle on a given sphere, the less is its distance from the centre.

2. Circles on the Sphere

We have seen that a plane cutting the sphere does so in a circle.

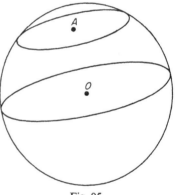

Fig. 95

Definitions. A circle whose plane passes through the centre of the sphere is called a *great* circle. Its radius is equal to that of the sphere.

A circle whose plane does not pass through the centre of the sphere is called a *small* circle. Its radius is less than that of the sphere.

Given a *great circle* lying in a plane α, the diameter perpendicular to α cuts the sphere in two points N, S called the *poles* of the great circle. The plane α is then called the *polar plane* of

N and of S. Any plane through the line NS cuts the sphere in a great circle whose plane is perpendicular to α.

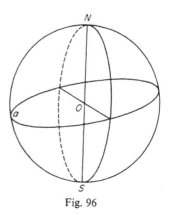

Fig. 96

It is sometimes convenient, for obvious reasons, to call the great circle in α the *equator* and N, S the *north* and *south* poles. The small circles in planes parallel to α are *circles of latitude* and the great circles through NS are *circles of longitude*.

Problems

1. If A lies in the polar plane of B, then B lies in the polar plane of A.
2. Every point of the sphere has a unique polar plane.
3. Two given points, not at the ends of a diameter, define a unique great circle and so a point on whose polar plane both lie.
4. Given three points A, B, U on a sphere (in general position on it) such that UA, UB both subtend a right angle at the centre O, then U is the pole of the great circle through A, B.

3. Spherical Triangles

It is not our aim to study in great detail the geometry of circles on a sphere, but one or two basic ideas may be helpful.

*It is assumed throughout this paragraph that all circles mentioned are **great** circles.*

Let Ω be a given sphere, and draw three diametral planes meeting it in great circles α, β, γ. There are two points common to each pair of circles and they lie at the ends of the diameter common to their planes. Suppose that β, γ meet in A, A'; γ, α meet in B, B'; α, β meet in C, C'. (In the diagram, α is taken as the plane of the paper.)

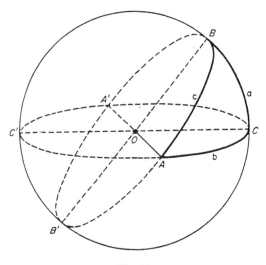

Fig. 97

Definition. The figure on a sphere bounded by arcs of three *great* circles is called a *spherical triangle*. (But see the last sentence of this paragraph.)

In the diagram, there are 8 triangles, ABC, $A'BC$, $AB'C$, ABC', $AB'C'$, $A'BC'$, $A'B'C$, $A'B'C'$, grouped in diametrically opposite pairs. We focus attention on one of them, say ABC.

For the purposes of calculations, mainly in spherical trigonometry with which we do not deal, the ground sphere Ω is taken to be of unit radius. Thus *a spherical triangle is defined by three* great *circles on a sphere of* unit *radius*.

The lengths of the sides of the triangle are the lengths of the arcs of the defining circles, say $BC = a$, $CA = b$, $AB = c$. Note that *the length of the arc BC is, by standard formula, equal to the radius of the circle $BCB'C'$ multiplied by the* radian measure *of the angle subtended by BC at the centre O*. It is in this sense that the side *BC* is often spoken of as an *angle* (in radian measure), namely the angle subtended by it at the centre when the sphere has unit radius. With this convention, a formula involving sin *a* or cos *a* has a clear meaning.

The *angles of the triangle* are defined to be the angles between the tangents at the vertices to the defining sides. For example, the angle *A* is defined to be the angle between the tangents at *A* to the circles *AB*, *AC*. Since these tangents are both perpendicular to *OA*, this angle is, by definition, the angle between the *planes* containing the circles *AB*, *AC*.

Finally, we remark that, given the vertices of a spherical triangle *ABC*, there still remains ambiguity, since there are *two* arcs of great circles joining, say, *B*, *C*. *It is agreed by convention that the triangle is so selected that its sides and angles are all less than π.*

4. The Polar Triangle of a Spherical Triangle

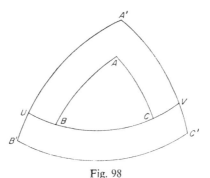

Fig. 98

Given a spherical triangle ABC whose sides a, b, c lie in planes α, β, γ, let A', B', C' be the poles of α, β, γ, selected in each case to lie on the same side of the plane as the corresponding vertex.

Definition. The spherical triangle with vertices A', B', C' is called the *polar triangle* of ABC.

(i) THE PROBLEM. It is required *to prove that ABC is the polar triangle of $A'B'C'$.*

THE DISCUSSION.

B' is the pole of $AC \Rightarrow B'A = \tfrac{1}{2}\pi$ (on unit sphere)
C' is the pole of $AB \Rightarrow C'A = \tfrac{1}{2}\pi$.

Thus
$$AB', AC' = \tfrac{1}{2}\pi.$$
$\Rightarrow A$ is the pole of the great circle $B'C'$.

Moreover,

A' on the same side of BC as $A \Rightarrow AA' < \tfrac{1}{2}\pi$
$\Rightarrow A$ on the same side of $B'C'$ as A'.

Hence, since similar results hold for B and C, the triangle ABC is the polar triangle of $A'B'C'$.

(ii) THE PROBLEM. It is required *to prove that, if the sides of the triangle ABC (in radian measure, for a unit sphere) are a, b, c and its angles A, B, C, and if a', b', c', A', B', C', are the corresponding magnitudes for the polar triangle $A'B'C'$, then*
$$a + A' = b + B' = c + C' = \pi,$$
$$a' + A = b' + B = c' + C = \pi.$$

THE DISCUSSION. Let BC meet $A'B'$ in U and $A'C'$ in V. Then
$$A' \text{ is the pole of } BC$$

$\Rightarrow A'U = A'V = \tfrac{1}{2}\pi$

$\Rightarrow UOV$ (where O is the centre of the sphere) is the angle between the planes $A'B'$, $A'C'$

$\Rightarrow UV = A'$ (in radian measure on the unit sphere).

But
$$B \text{ is the pole of } A'C'$$
$$\Rightarrow BV = \tfrac{1}{2}\pi;$$
$$C \text{ is the pole of } A'B'$$
$$\Rightarrow CU = \tfrac{1}{2}\pi.$$

Then
$$BV + CU = \pi$$
$$\Rightarrow (BC + CV) + CU = \pi$$
$$\Rightarrow BC + (CV + CU) = \pi$$
$$\Rightarrow BC + UV = \pi$$
$$\Rightarrow a + A' = \pi.$$

The results $b + B' = \pi$, $c + C' = \pi$ follow similarly. The other three formulae follow at once since ABC is the polar triangle of $A'B'C'$, so that rôles can be reversed.

Theorems

1. Two spherical triangles are congruent which have
   (i) three sides equal,
   (ii) two sides and the included angles equal,
   (iii) three angles equal.
2. If a spherical triangle has two sides equal, the corresponding angles are also equal.

Problems

1. The angles of a spherical triangle ABC, on a unit sphere, are all $\tfrac{1}{2}\pi$. Prove that the sides are all $\tfrac{1}{2}\pi$.

U is the point of BC produced such that $CU = \tfrac{1}{4}\pi$; V is the point of CB produced such that $CV = \tfrac{1}{4}\pi$. Prove that UV is a diameter of the sphere, and that a great circle can be drawn through U, V bisecting AB at P and AC at Q.

Prove that the triangles *BVP*, *APQ*, *CQU* are congruent, and deduce that $PQ = \frac{1}{3}\pi$.

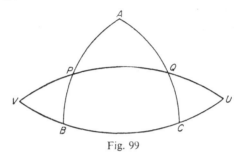

Fig. 99

2. In a spherical triangle *ABC* it is given that $AB = AC$, and *U* is the middle point of *BC*. Prove that $AU \perp BC$.

3. A spherical triangle has all its sides equal. Prove that its polar triangle also has all its sides equal.

5. Area on a Sphere

(i) The area of a lune

Definition. A *lune* on a sphere is that portion of the surface which is cut off between two diametral planes. (See the diagram, Fig. 100.)

If the angle between the two planes is θ, then we may call θ the *angle of the lune*.

As a matter of simple proportion, the area of the lune bears to the whole sphere the same ratio as θ (in *radian* measure) bears to 2π. Further, it is known that the area of a sphere of radius a is $4\pi a^2$, and so the *area of the lune* is $(\theta/2\pi)4\pi a^2$, or

$$2\theta a^2.$$

In particular, *the area of a lune of angle θ on a unit sphere is 2θ.*

Fig. 100

(ii) The area of a spherical triangle

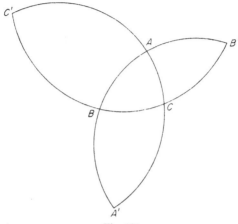

Fig. 101

THE PROBLEM. It is required *to prove that the area of a spherical triangle (on a unit sphere) whose angles in radian measure are $\hat{A}$, $\hat{B}$, $\hat{C}$ is given by the formula:*

$$area = A + B + C - \pi.$$

THE DISCUSSION. Let the other ends of the diameters through the vertices A, B, C of the given triangle be A', B', C'. Then (Fig. 101) $ABA'C$, $BCB'A$, $CAC'B$ are lunes of angles A, B, C so that the sum of their areas in $2(\hat{A} + \hat{B} + \hat{C})$.

Refer now to Fig. 97 (p. 134). The sum of the three lunes is the sum of the areas of the spherical triangles

$$(ABC + A'BC) + (ABC + AB'C) + (ABC + ABC')$$
$$= (ABC + A'BC + AB'C + ABC') + 2ABC.$$

But $\triangle AB'C \equiv \triangle A'BC'$, and so the sum in brackets is (rearranging)

$$ACB + CA'B + A'C'B + C'AB,$$

which, by inspection, is the sum of the four triangles in the "upper" hemisphere of the diagram. Hence

$$2(\hat{A} + \hat{B} + \hat{C}) = \text{hemisphere} + 2\triangle ABC$$
$$= 2\pi + 2\triangle ABC,$$

so that

$$\triangle ABC = \hat{A} + \hat{B} + \hat{C} - \pi.$$

COROLLARY. Since the area is necessarily positive, *the sum of the angles of a spherical triangle is greater than π*.

Definitions. The quantity $\hat{A} + \hat{B} + \hat{C} - \pi$ is called the *spherical excess* of the spherical triangle *ABC*.

Problems

1. If *OA, OB, OC* are three mutually perpendicular radii of a sphere of unit radius, the area of the spherical triangle *ABC* is $\tfrac{1}{2}\pi$.
2. Prove that, in a spherical triangle *ABC*,

$$\hat{A} + \hat{B} + \hat{C} > \pi.$$

By considering the corresponding result for the polar triangles, deduce that

$$a + b + c < 2\pi.$$

6. The Right Circular Cone

Definitions. Given a point *O* and a curve not lying in a plane through it, the surface traced out (*generated*) by lines passing through *O* and a variable point of the curve is called a *cone* of vertex *O*, the variable lines being called *generators*.

When the curve is a circle whose centre *D* is the foot of the perpendicular from *O* to the plane of the circle, the cone is said to be *right circular*. The line *OD* is called the *axis* of the cone.

By congruent triangles, the angle between the axis and a generator has a constant value, known as the *semi vertical angle* of the

The Sphere 141

cone. If $OD = h$ and the radius of the circle is a, then the semi-vertical angle θ satisfies the relation

$$a = h \tan \theta.$$

In some contexts, the cone is regarded as truncated to lie

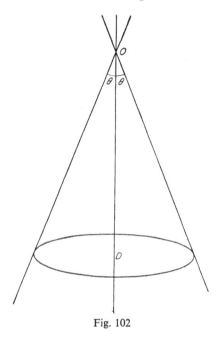

Fig. 102

between O and the plane of the defining circle. Then OD is called the *height*, the circle is called the *base* and the length of the segment of a generator intercepted between O and the base is called the *slant height*.

Theorems
1. A plane through O cuts the cone *either* at O only *or* in two generators. Exceptionally, the plane may cut the cone in a single generator, when

its plane cuts the base in a straight line touching the circle there. Such a plane is called a *tangent plane* to the cone.

2. Every plane parallel to the base cuts the cone in a circle. Each tangent line to the cone cuts such a plane in a line which is a tangent to the circle of section.

3. The tangent lines to a sphere which pass through a point O outside the sphere are the generators of a right circular cone of vertex O. The points common to the cone and the sphere lie on a circle.

7. The Right Circular Cylinder

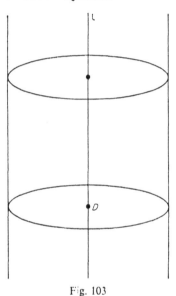

Fig. 103

Definitions. Given a line l and a curve not lying in a plane through it, the surface *generated* by lines parallel to l and passing through a variable point of the curve is called a *cylinder* of which the variable lines are *generators*.

When the curve is a circle whose centre D lies on l and whose plane is perpendicular to *l*, the cylinder is said to be *right circular*. The line *l* is the *axis* of the cylinder.

Theorems
1. The planes perpendicular to the axis cut the cylinder in circles of constant radius.
2. The tangent lines to a sphere which are parallel to a given direction are the generators of a right circular cylinder. The points common to the cylinder and the sphere lie on a circle.

Problems
1. A right circular cone of base radius a and height h is cut by a right circular cylinder of radius $\frac{1}{2}a$ whose axis coincides with that of the cone. Prove that the points common to the cylinder and the cone (extended "beyond its vertex") lie on one or other of two equal circles.
2. A right circular cone has vertex A, axis AB and vertical angle $60°$; another right circular cone has vertex B, axis BA and vertical angle $30°$. Prove that the points common to the cones lie on one or other of two circles whose radii are in the ratio 1:2.
3. Prove that the radius of the largest sphere that can be inscribed in a right circular cone of height h and base radius a is

$$a\{\sqrt{(a^2 + b^2)} - a\}/h.$$

4. Prove that the points common to a right circular cone and a sphere whose centre is on the axis lie on one or other of two circles.

Prove the corresponding result when the cone is replaced by a right circular cylinder.
5. Prove that a right circular cone and a right circular cylinder having the same axis meet in the points of two circles (one on either side of the vertex of the cone).

ELEVEN
The Nature of Space

THE THEOREMS outlined in Chapter 1 have a dual purpose: they form a basis for the study of geometrical relationships founded on logical argument, and, even more fundamentally, they seek to describe the structure of space itself in so far as it is concerned with such matters as size and relative position. The ideas of point, line, length, angle, are undoubtedly abstract, but they are designed to agree as closely as possible with the physical world of sight and touch.

There are certain relationships of particular importance which may usefully be emphasised here. It is probable that they will present little that is new, but the attempt to classify them helps towards clearer thinking.

For simplicity of statement we confine ourselves mainly to geometry in a plane.

1. Translation

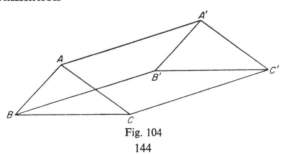

Fig. 104

The Nature of Space

Let *ABC* be a given triangle. It may be moved bodily in any given direction (without "rotation", which we shall discuss later) to a new position $A'B'C'$ by moving each of the vertices an agreed distance in that direction; thus

$$AA' \| BB' \| CC' \text{ and } AA' = BB' = CC'.$$

Definition. Such a movement from position *ABC* to position $A'B'C'$ is called a *translation*.

It is an immediate consequence of the theorems of Chapter 1 that

$$B'C' = BC, C'A' = CA, A'B' = AB,$$

so that

$$\triangle A'B'C' \equiv \triangle ABC;$$

that is, *the triangle in the new position under a translation is congruent to the triangle in the old position*. The point to be emphasized, though, is that all this arises from our instinctive belief that space itself has such a property—space, so to speak, does not "crinkle".

2. Rotation

Take the triangle *ABC* as before, and select a point *O* in its plane. Rotate the triangle about *O* through an angle θ to a position $A'B'C'$: then

$$OA' = OA, OB' = OB, OC' = OC,$$
$$\angle AOA' = \angle BOB' = \angle COC' = \theta.$$

Definition. The movement from position *ABC* to position $A'B'C'$ is called a *rotation* with *centre O*.

When a position $A'B'C'$ is *known* to be obtainable by rotation from a position *ABC*, the centre *O* is easily located: *it lies at the intersection of the perpendicular bisectors of AA' and BB'*.

Here, again, the result of congruence holds:

$$\triangle A'B'C' \equiv \triangle ABC.$$

146 *Deductive Geometry*

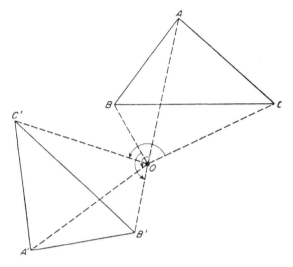

Fig. 105

3. Direct and Inverse Congruence

Experience with left-handed and right-handed gloves will already have convinced the reader that there are two kinds of congruence: one in which a triangle $A'B'C'$ can be moved continuously to a congruent triangle ABC so as to lie on it point for point (A' on A, B' on B, C' on C), and the other in which such superposition is not possible until the triangle has been "turned over" first.

The two cases may be called *direct congruence* and *inverse congruence* respectively.

The "turning over" of the triangle may be described more scientifically as a rotation of 180° about a line *l* lying in its plane.

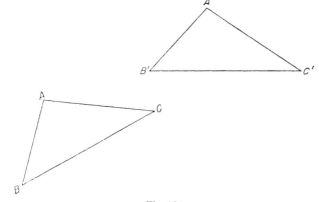

Fig. 106

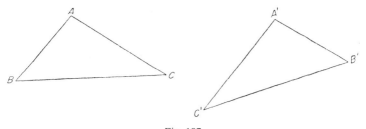

Fig. 107

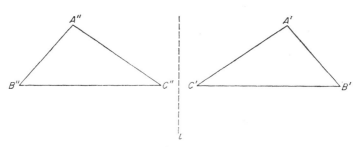

Fig. 108

148   *Deductive Geometry*

Note. A point may be regarded as "going round" the sides of a triangle if it is conceived as moving along the sides $\overrightarrow{BC}$, $\overrightarrow{CA}$, $\overrightarrow{AB}$ in the sense implied by the arrows.

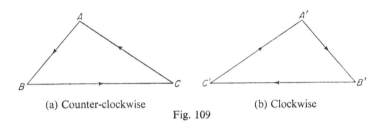

(a) Counter-clockwise   (b) Clockwise

Fig. 109

In Fig. 109 (a) the sense may be called *counterclockwise*, and in Fig. 109 (b) it may be called *clockwise*. Then the congruence

$$\triangle ABC \equiv \triangle A'B'C'$$

(with A corresponding to A', B to B' and C to C') is direct if both senses are counterclockwise or both clockwise, and inverse if the two are opposite.

4. The Rotation of a Configuration

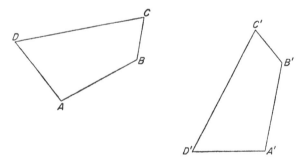

Fig. 110

The Nature of Space

It is a problem of interest to decide whether two congruent configurations can be rotated the one to coincide with the other. This will certainly not be possible unless the congruence is *direct*.

On the other hand, if the congruence *is* direct, then it is sufficient to consider only two corresponding lines, say AB, $A'B'$; for if $A'B'$ is brought to coincidence with AB, the other points will automatically fall into position.

Suppose, then, that we are given two equal lines AB, $A'B'$. If they *can* be brought to coincidence by rotation, the centre O, being equidistant from A, A' and from B, B', must lie on the perpendicular bisectors of AA' and BB'.

Let these perpendicular bisectors be constructed, as in the diagram. [We pass over a point of instrinsic difficulty about the relative senses in which the angles OAB, $OA'B'$ turn. The diagram is correct, but it is not easy to prove that it *must* be. For a discussion of the problems involved, see E. A. Maxwell, *Fallacies in Mathematics*, Cambridge University Press (1959) p. 34. The significance of the dotted lines in the diagram will appear later.]

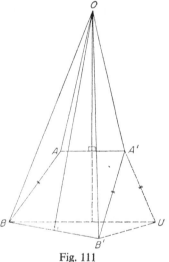

Fig. 111

Now

O on perpendicular bisectors of AA', BB'
$\Rightarrow OA = OA', OB = OB'$

and
$$OA = OA', OB = OB', AB = A'B'$$
$$\Rightarrow \triangle OAB \equiv \triangle OA'B'$$
$$\Rightarrow \angle BOA = \angle B'OA'$$
$$\Rightarrow \angle BOA + \angle AOB' = \angle AOB' + \angle B'OA'$$
$$\Rightarrow \angle BOB' = \angle AOA'.$$

Hence rotation about O through an angle AOA' (or BOB') carries A to A' and B to B', so that AB can be *rotated*† to the position $A'B'$.

Note. The difficulty in this proof is to make sure that the triangles OAB, $OA'B'$ lie on the correct sides of OA, OA' for the additions

"$\angle BOA + \angle AOB' = \angle AOB' + \angle B'OA'$
$\Rightarrow \angle BOB' = \angle AOA'$"

to be legitimate. If, however, $\triangle OA'B'$ is rotated about OA' over to the position $OA'U$, then it can be shown that $BU \| AA'$, the figure $AA'UB$ being symmetrical about the perpendicular bisector of AA'. Thus U lies "out from" OA' just as B lies "out from" OA; and B' is therefore in the position shown. But this is hard for a reader at the level of study envisaged and may perhaps be accepted for the present.

5. Expansion; Homothetic figures

Let ABC be a given triangle and O a point in its plane (inside or outside the triangle). Take points A', B', C' on $\overrightarrow{OA}$, $\overrightarrow{OB}$, $\overrightarrow{OC}$ so that
$$OA' = kOA, OB' = kOB, OC' = kOC.$$

Then it is an immediate exercise in similar triangles that
$$\triangle A'B'C' \sim \triangle ABC,$$

† If, exceptionally, $AB \| A'B'$, the argument breaks down.

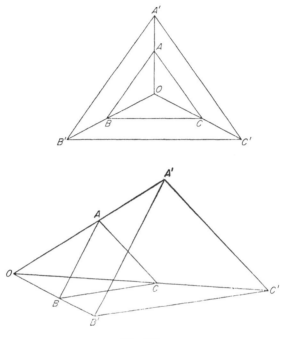

Fig. 112

with

$$B'C' \| BC,\ C'A' \| CA,\ A'B' \| AB,$$
$$B'C' = kBC,\ C'A' = kCA,\ A'B' = kAB.$$

The triangle $A'B'C'$ is an *expansion* ($k > 1$) or *contraction* ($k < 1$) of $\triangle ABC$.

Definition. Two figures related in this way are said to be *homothetic* or *similar and similarly situated*. They may be called *directly* similar in that the senses of description $\vec{BC}$, $\vec{CA}$, $\vec{AB}$ and $\vec{B'C'}$, $\vec{C'A'}$, $\vec{A'B'}$ are the same.

Suppose, next, that A', B', C' are taken on $\vec{AO}$, $\vec{BO}$, $\vec{CO}$ so that
$$OA' = kAO, \quad OB' = kBO, \quad OC' = kCO.$$

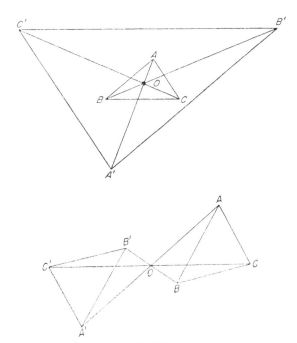

Fig. 113

Then, once again,
$$\triangle A'B'C' \sim \triangle ABC,$$
with
$$B'C' \| BC,\ C'A' \| CA,\ A'B' \| AB,$$
$$B'C' = kBC,\ C'A' = kCA,\ A'B' = kAB.$$

The triangles may be called *inversely similar* in that the senses of description $\vec{BC}$, $\vec{CA}$, $\vec{AB}$ and $\vec{B'C'}$, $\vec{C'A'}$, $\vec{A'B'}$ are opposite.

6. Symmetry

Definition. A figure is said to be *symmetric about a line l* if corresponding to each point P of the figure there is another point Q of the figure such that the line PQ is perpendicular to l and bisected by l. The line l is called the *axis of symmetry*.

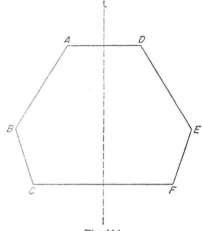

Fig. 114

Typical symmetrical figures are the rectangle, the isosceles triangle, the circle, as indicated in the diagram (Fig. 115).

A figure may have more than one axis of symmetry. For

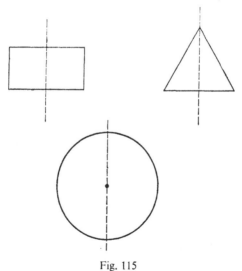

Fig. 115

example, a rectangle has two, an equilateral triangle three and a square four. A circle is symmetrical about every diameter.

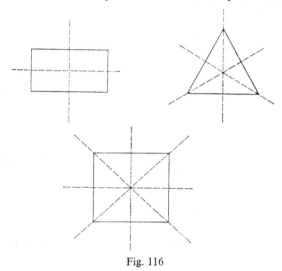

Fig. 116

A figure may also be *symmetric about a point O* (the *centre of symmetry*), when corresponding to each point P of the figure there is another point Q of the figure such that O is the middle point of PQ. Thus a rectangle is symmetric about its centre, and a circle is symmetric about its centre.

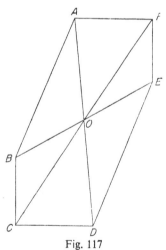

Fig. 117

Theorem

1. If a figure is symmetric about a line l and a line m, it is also symmetric about the point $O = l \cap m$.

TWELVE
Transformations

THERE ARE many problems in geometry where a figure F can be brought into close relationship with another figure F' which, at first sight, is very dissimilar. The advantages are twofold: on the one hand, an unexpected unity can be brought to the subject when it is realized that apparently distinct configurations are, basically, just different aspects of one another; on the other hand, it often happens that the properties of one aspect are particularly simple, or particularly familiar, and then the corresponding properties of a more complicated alternative can be "read off" with comparative ease.

1. Orthogonal Projection

Definition. The foot of the perpendicular from a point A on to a given plane π is known as the *orthogonal projection* of A on π.

If A, B are two points whose orthogonal projections on π are A', B', then it follows directly from the work of Chapter 6 that the orthogonal projection of every point of AB lies on $A'B'$. That is, if $C \in AB$, then $C' \in A'B'$.

More generally, if a figure F lies in a plane α, then the corresponding figure F' in π obtained by taking the orthogonal projections of all the points of F is called the *orthogonal projection* of F on π.

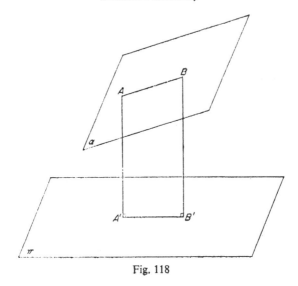

Fig. 118

Theorems

[The points A', B', C', ... are the orthogonal projections on π of points A, B, C, ... in α.]

1. A, B, C collinear $\Leftrightarrow$ A', B', C' collinear.
2. $AB/BC = A'B'/B'C'$.
3. If $L = AB \cap CD$, then $L' = A'B' \cap C'D'$.
4. If $AB \| CD$, then $A'B' \| C'D'$.
5. If $ABCD$ is a parallelogram, then $A'B'C'D'$ is a parallelogram.
6. A harmonic range (AB, CD) projects orthogonally into a harmonic range $(A'B', C'D')$.
7. Let the planes α, π meet at an angle θ, and denote by l their line of intersection. Then

   (i) the length of the projection $A'B'$ of a line AB parallel to l is unchanged, so that $A'B' = AB$;

   (ii) the length of the projection $C'D'$ of a line CD perpendicular to l is reduced in the ratio $\cos \theta$, so that $C'D' = CD \cos \theta$.
8. If a triangle ABC in the plane α is projected into a triangle $A'B'C'$ in π, then area $A'B'C'$ = area $ABC \times \cos \theta$.

Problems

1. A line AB lies in the plane α and makes an angle ϕ with the line of intersection of α, π. Its orthogonal projection on π is $A'B'$, and θ is the angle between the planes α, π. Prove that

$$\left(\frac{A'B'}{AB}\right)^2 = \cos^2 \phi + \sin^2 \phi \cos^2 \theta.$$

Prove also that, if ψ is the angle between $A'B'$ and the line of intersection of the planes, then

$$\tan \psi = \tan \phi \cos \theta.$$

2. Prove that, in general, if $ABCD$ is a rectangle, then $A'B'C'D'$ is not a rectangle.

3. Show how to project (a) a rectangle, (b) a rhombus orthogonally into a square.

2. Conical Projection

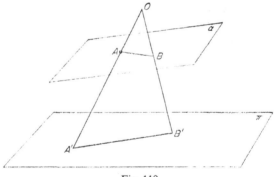

Fig. 119

Definition. Let O be a given point (the *vertex of projection*) and π a given plane (the *plane of projection*). If A is any point in general position in space, then the point where the line OA meets π is called the *projection* (or *conical projection*) of A on π; it is usually denoted by the name A'.

If A, B are two given points and A', B' their projections, then the projection of any point of AB lies on $A'B'$.

(i) The vanishing plane

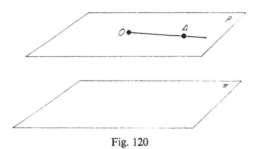

Fig. 120

Definition. The plane ρ through O parallel to π is called the *vanishing plane* for the projection. The reason for the name is simple: if A is any point in the plane, the line OA cannot meet π and so the projection A' is non-existent.

For a figure lying in a general plane α, the intersection of α

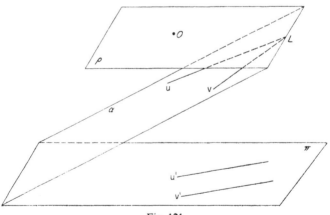

Fig. 121

Transformations

with the vanishing plane ρ is called the *vanishing line in the plane α for the projection on π by means of the vertex O.*

In particular, *two lines u, v of α which meet on the vanishing line project into two* parallel *lines u', v' in π*:—

For an intersection L' of u', v' would arise from the intersection L of u, v. But, by definition of the plane ρ, L' cannot exist, and so u', v' are parallel.

(ii) THE PROBLEM. It is required *to project a given quadrangle $ABCD$ into a parallelogram $A'B'C'D'$*.

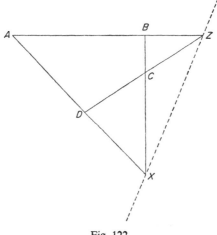

Fig. 122

THE DISCUSSION. Write $X = BC \cap AD$, $Z = AB \cap CD$.

Through XZ draw any plane ρ, and select any point O in it. Take any plane π parallel to ρ. Then, by what we have just done,

$$B'C' \| A'D',\ A'B' \| C'D',$$

so that $A'B'C'D'$ is a parallelogram.

160   *Deductive Geometry*

(iii) THE PROBLEM. Given two lines u, v, intersecting in A, lying in the plane α and meeting the vanishing line in points U, V, it is required *to prove that the angle between u' and v' is equal to $\angle UOV$.*

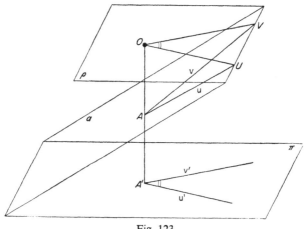

Fig. 123

THE DISCUSSION. By definition of conical projection, OAA', u, u' are coplanar, so that

$$OU \| u'$$

and OAA', v, v' are coplanar, so that

$$OV \| v'.$$

Hence the angle between u', v' is $\angle UOV$.

(iv) To project a given quadrangle into a square

THE PROBLEM. Given a quadrangle $ABCD$, it is required *to find a vertex and plane of projection such that $A'B'C'D'$ is a square.*

THE DISCUSSION. A parallelogram is a square if one angle, say $D'A'B'$, is a right angle and if the angle $B'U'C'$ between the diagonals is also a right angle.

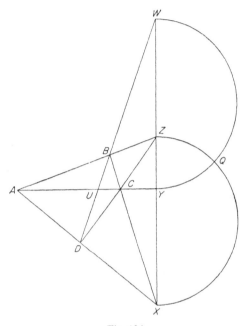

Fig. 124

Let $X = BC \cap AD$, $Y = CA \cap BD$, $Z = AB \cap CD$, $W = BD \cap YZ$. Following on (iii), take XYZ as vanishing line. In order to get the right angles, let Q be one of the points of intersection of the circles on XZ, YW as diameters. Rotate the lines QX, QY, QZ, QW about XYZ out of the plane α of $ABCD$ so that Q now assumes a position O, the plane OXW being the vanishing plane ρ. Any plane π parallel to ρ can then be taken as the plane of projection.

The proof is immediate, since
$$\angle D'A'B' = XOZ = \angle XQZ = 90°,$$
and
$$\angle B'U'C' = \angle WOY = \angle WQY = 90°.$$

Theorem

1. A harmonic range (AB, CD) projects conically into a harmonic range $(A'B', C'D')$.

If, however, A is on the vanishing line for the projection, then B' is the middle point of $C'D'$.

Problems

1. Prove by the conical projection of a quadrangle $ABCD$, with the side XZ of the diagonal triangle as vanishing line, that the diagonals of a parallelogram $A'B'C'D'$ bisect each other.

2. (AB, UP) and (AC, VQ) are harmonic ranges. Prove that UV, BC meet on PQ.

By conical projection of this figure with PQ as vanishing line, prove that the line joining the middle points of the sides of a triangle is parallel to the base.

By conical projection making $UVCB$ a parallelogram, prove that the line joining the middle points of one pair of opposite sides of a parallelogram is parallel to the other pair.

3. Two triangles ABC, PQR are so related (in *perspective*) that AP, BQ, CR have a common point O; $L = BC \cap QR$, $M = CA \cap RP$, $N = AB \cap PQ$. Prove that, in a projection with MN as vanishing line, the projections $B'C'$, $Q'R'$ of BC, QR are parallel. Deduce that $L \in MN$ (Theorem of *Desargues*).

4. A, B, C and P, Q, R are two sets of collinear points on distinct lines; $L = BR \cap CQ$, $M = CP \cap AR$, $N = AQ \cap BP$. Prove that, in a projection with MN as vanishing line, the projections $B'R'$, $C'Q'$ of BR, CQ are parallel. Deduce that $L \in MN$ (Theorem of *Pappus*).

5. Establish the equivalence of the following theorems:

   (a) $ABCD$, $A'B'C'D'$ are two quadrangles so related that $BC \cap AD = B'C' \cap A'D' = X$, $AB \cap CD = A'B' \cap C'D' = Z$.

   Then, if $BD \cap B'D' \in XZ$, it follows that $AC \cap A'C' \in XZ$.

   (b) $ABCD$, $A'B'C'D'$ are two parallelograms so related that $AB\|CD\|A'B'\|C'D'$ and $AD\|BC\|A'B'\|C'D'$.

   Then, if $BD\|B'D'$, it follows that $AC\|A'C'$.

   Give independent proofs of each of the results.

6. $ABCD$ is a quadrangle; $X = BC \cap AD$, $Y = CA \cap BD$, $Z = AB \cap CD$. A line through Y meets AB in P, CD in R; another line through Y meets BC in Q, AD in S. By means of a projection with XZ as vanishing line, prove that $PQ \cap RS \in XZ$, $PS \cap QR \in XZ$.

7. ABC is a triangle. Parallel lines AP, BQ, CR meet BC, CA, AB in P, Q, R; $L = QR \cap BC$, $M = RP \cap CA$, $N = PQ \cap AB$. Prove that $L \in MN$.

8. A, B, O are three non-collinear points, and parallel lines AU, BV are drawn; $P = OA \cap BV$, $Q = OB \cap AU$, R is the point where the line through O parallel to AU and BV meets AB; $L = QR \cap BV$, $M = RP \cap AU$, $N = PQ \cap AB$. Prove that $L \in MN$.

9. A line LMN meets the sides BC, CA, AB of $\triangle ABC$ in L, M, N. Points A', B', C' are chosen on BC, CA, AB so that harm. $(A'L, BC)$, harm. $(B'M, CA)$, harm. $(C'N, AB)$. By means of a projection with LMN as vanishing line, prove that AA', BB', CC' are concurrent.

10. Establish the following interpretation of the problem stated in converse in question 9:

$B'CC'B$ is a parallelogram whose diagonals BC, $B'C'$ meet in L; a line through L meets $B'C$ in M and BC' in N. If the line is chosen so that $NB = BC'$, then $MC = CB'$.

3. Inversion

Definition. Recall, first, the definition of inverse points. Given a circle Ω of centre O and radius a, the *inverse* of a point A with respect to Ω is the point A' (on the same side of O as A) such that $OA \cdot OA' = a^2$.

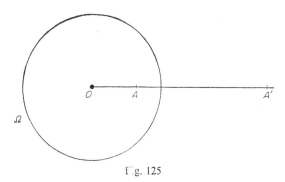

Fig. 125

In many problems the actual circle Ω is irrelevent so long as its centre O is known. We then speak of *inversion with respect to O*.

If A moves on some such curve as a straight line or a circle, the locus of A' is called the *inverse curve* of the given curve with respect to Ω.

(i) The magnification theorem

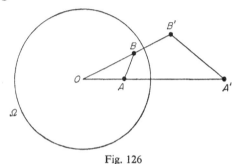

Fig. 126

THE PROBLEM. Let A, B be two given points, with inverses A', B'. The *magnification* of AB under the inversion may be defined as the ratio $A'B'/AB$. It is required to *prove that*

$$\frac{A'B'}{AB} = \frac{OB'}{OA} = \frac{OA'}{OB}.$$

THE DISCUSSION.
$$OA \cdot OA' = a^2 = OB \cdot OB'$$
$$\Rightarrow \frac{OA}{OB} = \frac{OB'}{OA'},$$

and

$$\left.\begin{array}{r}\dfrac{OA}{OB} = \dfrac{OB'}{OA'} \\ \angle AOB = \angle B'OA'\end{array}\right\} \Rightarrow \triangle AOB \sim \triangle B'OA'$$

$$\Rightarrow \frac{AB}{B'A'} = \frac{OA}{OB'} = \frac{OB}{OA'}.$$

Thus

$$A'B' = \frac{OB'}{OA} \cdot AB = \frac{OA'}{OB} \cdot AB.$$

(ii) We can now find the inverses of straight lines and circles:

(a) *To prove that the inverse of a straight line not through O is a circle through O.*

Let l be the given line, O the centre of the circle of inversion Ω, A the foot of the perpendicular from O to l and A' the inverse of A.

Take an arbitrary point P on l, and let P' be its inverse. Then

$OA \cdot OA' = a^2 = OP \cdot OP'$
$\Rightarrow A, A', P, P'$ concyclic,

and

$\angle A'AP = 90° \Rightarrow A'P$ subtends a right
$\qquad$ angle on the circle $AA'P'P$
$\qquad \Rightarrow \angle A'P'P = 90°$
$\qquad \Rightarrow \angle OP'A' = 90°.$

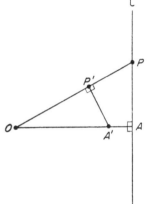

Fig. 127

But O, A' are fixed points, and so P' lies on the circle on OA' as diameter.

(b) *To prove that the inverse of a circle through O is a straight line not through O.*

Let m be the given circle, O the centre of the circle of inversion Ω, A the other end of the diameter of m through O, and A' the inverse of A.

Take an arbitrary point P on m, and let P' be its inverse. Then

$$OA \cdot OA' = a^2 = OP \cdot OP'$$
$$\Rightarrow A, A', P, P' \text{ concyclic,}$$

166 *Deductive Geometry*

and

$$OA \text{ diameter} \Rightarrow \angle OPA = 90°$$
$$\Rightarrow \angle OA'P' = 90°.$$

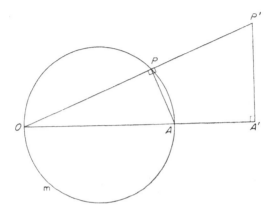

Fig. 128

But O, A' are fixed points, and so P' lies on the straight line through A' perpendicular to OA'.

Problems

1. Two circles cut orthogonally at O. Prove that their inverses with respect to O are two perpendicular lines.

2. Two circles touch at O. Prove that their inverses with respect to O are two parallel lines.

3. Two lines l, m are parallel and $O \in l$. Prove that the inverse with respect to O of l is l itself, and that the inverse of m is a circle touching l at O.

(c) *To prove that the inverse of a circle not through O is a circle not through O.*

Let m be the given circle, O the centre of the circle of inversion Ω, B the inverse of O with respect to the circle m, and B' the inverse of B with respect to Ω.

Transformations

What we shall prove is that the inverse of m is a circle m' whose centre is B', so that the method of proof locates the centre of m' at the same time.

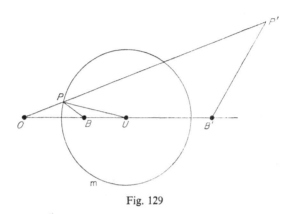

Fig. 129

Take an arbitrary point P on m, and let P' be its inverse with respect to Ω. Then, by the magnification theorem,

$$B'P' = \frac{OB'}{OP} \cdot BP$$
$$= \frac{BP}{OP} \cdot OB'$$

Suppose now that U is the centre of the circle m. Then
O, B inverse with respect to m

$$\Rightarrow UP^2 = UB \cdot UO \Rightarrow UP/UB = UO/UP$$
$$\Rightarrow \triangle UPB \sim \triangle UOP$$
$$\Rightarrow \frac{PB}{OP} = \frac{UP}{UO}$$

Hence

$$B'P' = \frac{UP}{UO} \cdot OB'$$
$$= \text{constant},$$

since UP is the radius of m and U, O, B' are fixed points, and so P' lies on the circle of centre B' and radius $B'P'$.

(iii) The angle between two curves

In order to make full use of the technique of inversion, we must give a short account of the angle between two curves at a common point. The basic ideas will be familiar from work on calculus.

(a) *The tangent at a point.*

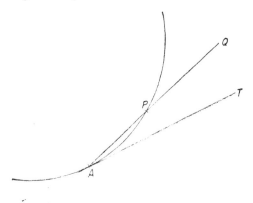

Fig. 130

Let A be a point on a given curve. We seek *to define the tangent to the curve at A*. Let P be any point of the curve, fairly near to A, and consider the line AQ through P. The point P may be supposed to approach more and more closely to A, and the line AQ will then (in normal cases) take up a limiting positive *AT* known as the *tangent at A* to the curve.

(b) *The angle between two curves.*

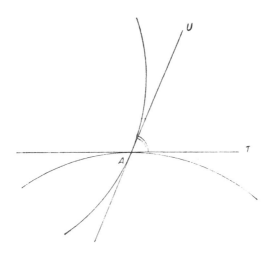

Fig. 131

If two curves cut at a point *A*, *the angle between the curves* at *A* is defined to be the angle between the tangents there.

(c) *The fundamental theorem:*

The angle between two curves is equal to the angle between their inverses.

Let *m*, *n* be two given curves intersecting at *A*. Take a point *P* on *m*, fairly near to *A*, and let *OP* meet *n* in *Q*. The inverses *P'*, *Q'* on the inverse curves *m'*, *n'* also lie on the line *OPQ*. Suppose, finally, that *A'* is **the** inverse of *A*; the two curves *m'*, *n'* then necessarily intersect **at** *A'*.

Now

$$OP \cdot OP' = OQ \cdot OQ' = OA \cdot OA'$$
$$\Rightarrow APP'A' \text{ cyclic and } AQQ'A' \text{ cyclic}$$
$$\Rightarrow \angle OAP = \angle OP'A', \angle OAQ = \angle OQ'A'$$
$$\Rightarrow \angle OAQ - \angle OAP = \angle OQ'A' - \angle OP'A'$$
$$\Rightarrow \angle PAQ = \angle P'A'Q'.$$

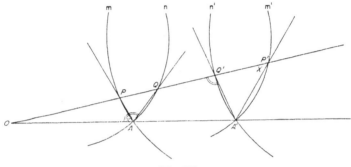

Fig. 132

Thus, however close P is to A (so that Q is also close to A), the angle between the *chords* $A'P'$ and $A'Q'$ is equal to the angle between the *chords* AP and AQ. In the limit, then, the angle between the *tangents* at A is equal to the angle between the *tangents* at A'.

INTERPRETATION. (a) Two curves which have the same tangent at A are said to *touch* at A. Their inverse curves will also touch at A—but note that the tangent AT at A will, in general, NOT invert into the tangent $A'T'$

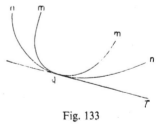

Fig. 133

Transformations

at A', since straight lines usually do not invert into straight lines but into circles.

When A, where the curves touch, is also the centre of inversion, more complicated problems arise. For straight lines and circles, the rules are:

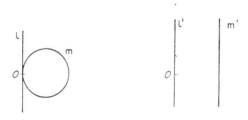

Fig. 134

(i) When a line l touches a circle m at O, the inverses are a line l' which is the same as l and a line m' which is parallel to l'.

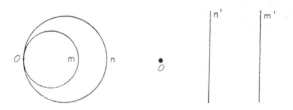

Fig. 135

(ii) When two circles m, n touch at O, the inverses are two *parallel* straight lines m', n'.

(b) Two curves are called *orthogonal* when the angles between them is a right angle. In particular, *when a line l is orthogonal to a circle m, then l is a diameter of m.*

172 *Deductive Geometry*

This interpretation for line and circle is very important in problems involving inversion.

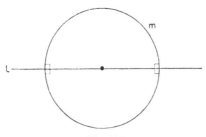

Fig. 136

Theorem

1. A circle and a diameter invert, with respect to a point on the circle, into "a diameter and a circle".

Problems

1. Two circles α, β touch at O. A circle Ω through O cuts them again at A, B respectively. Verify (what is in any case obvious at O) that the circles α, Ω cut at A at the same angle that the circles β, Ω cut at B. (Invert with respect to O.)

2. A point O lies on a circle having BC as diameter and A is a point on BC. By inversion with respect to O, prove that the circles OAB, OAC cut orthogonally.

3. Establish the equivalence of the two following theorems:

   (a) Two circles $OAPX$, $OBQX$ meet in O, X and PXQ, AOB are straight lines. Then $AP \| BQ$.

   (b) Two straight lines APX, BQX meet in X and O is a point on AB; the points O, P, X, Q are concyclic. Then the circles OAP, OBQ touch at O.

4. Given a circle l and two points A, O not on it, prove that a unique circle can be drawn through A, O to cut l orthogonally.

5. A, B, C are three collinear points and O a point not on the line ABC. Prove that there is a common point H to each of the circles passing through O and one of the points A, B, C and cutting orthogonally the circle through O and the two others of the points A, B, C.

6. Establish the equivalence of the following theorems:

(a) Given three non-collinear points A, B, C, the circles on AB, AC as diameters meet on BC.

(b) Given three non-collinear points A, B, C, the line through B perpendicular to AB meets the line through C perpendicular to AC on $\odot ABC$.

(c) Given three non-collinear points A, B, C, the line through A perpendicular to AB cuts the circle through A, C cutting $\odot ABC$ orthogonally in a point on BC.

7. Four points A, B, C, O lie on a given circle. A circle through B, C touches a circle through A, O at a point P. Prove that, as the touching circles vary, the locus of P is a circle.

8. Given three fixed points O, A, B and a variable point P on *either* the line AB *or* a fixed circle through A, B, prove that the circles OPA, OPB cut at a constant angle.

9. Establish the equivalence of the following theorems:

(a) Two circles p, q touch at O and two circles r, s touch at O. If one of the pairs of circles (p, r), (p, s), (q, s), (q, r) is orthogonal, so are the other three pairs.

(b) If one angle of a parallelogram is a right angle, so are the other three.

10. In question 9(a), the further intersections of the four pairs of circles, in that order, are A, B, C, D. Prove that, if a circle can be drawn to touch each of the circles p, q, r, s, then the circles OAC, OBD are orthogonal.

Index

(*Results Assumed Known*, pp. 6–13, *are not listed in detail.*)

Altitude, of triangle, 17
  of tetrahedron, 114
Angle between curves, 168
  line and plane, 120
  lines, 93
  planes, 121
Angle bisectors, harmonic property, 72
Angle, trihedral, 124
Apollonius, circle of, 50
Area, of spherical triangle, 139

Bisection theorems for harmonic pencil, 73

Centroid, of triangle, 14
  of tetrahedron, 110
Ceva, 53
Circles on sphere, 132
Circumcentre, of triangle, 16
  of tetrahedron, 118
Circumcircle, 16
Coaxal circles, 49
Cone, 140
Congruence, direct and inverse, 146
Conical projection, 157
Conjugate chords, 83
  points. 78
Cross-ratio 57
Curve, tangent, 168
Curves, orthogonal, 171
Cylinder, 142

Desargues, 56, 92, 162
Direct common tangents, 37
Duality, 66

Escribed circles, 19
Euler line, 18, 26
  theorem for convex body, stated, 129
Expansion, 150

Great circle, 132

Harmonic pencil, range, 58
  on circle, 85
  test $OA^2 = OP \cdot OQ$, 76
Harm. (AB, PQ), 61
Homothetic figures, 150

Incentre, 19
Incidence, propositions, 90
Inverse, 46
Inversion, 163

Limiting points, 42, 47
Lune, 138

Magnification theorem for inversion, 164
Medians, 14
Menelaus, 52
Mirror image, 33

Index

Nine-points circle, 22, 24
Normal, 94, 101
Notation used, 2–6

Orthocentre, of triangle, 17
   of tetrahedron, 116
Orthogonal circles, 44
   curves, 171
   projection, 155
   tetrahedron, 115

Pappus, 162
Parallel lines, 91
   planes, 97
Parallelepiped, 107
Perspective, 92
Platonic solids, 126
Pole and polar for triangle, 71
   for circle, 78
   for great circle on sphere, 133
Power of point with respect to circle, 43
Projection, conical, 157
   orthogonal, 155
Ptolemy, 27

Quadrangle, 64
   projected to parallelogram, 159
   square, 160

Radical axis, 40, 47

Regular solids, 126
Rotation, 145, 148

Self-conjugate (self-polar) triangle, 83
Sense on a line, 51
Similitude, 36, 39
Simson line, 30, 32
Skew lines, 91
   common perpendicular, 104
Slope, line of greatest, 122
Solids (regular, Platonic), 126
Sphere, 130
   circles on, 132
   tangent plane, 131
Spherical excess, 140
Spherical triangle, 134
   area, 139
   polar triangle of, 135
Symmetry, 153

Tangent to curve, 168
   plane to sphere, 131
Tetrahedron, 109
   orthocentre, 114
   orthogonal, 115
Three perpendiculars, theorem, 103
Translation, 144
Transversal, 92
Transverse common tangents, 37
Triangle, spherical, 134

Vanishing line, 159
   plane, 158

A CATALOG OF SELECTED
DOVER BOOKS
IN SCIENCE AND MATHEMATICS

CATALOG OF DOVER BOOKS

Astronomy

CHARIOTS FOR APOLLO: The NASA History of Manned Lunar Spacecraft to 1969, Courtney G. Brooks, James M. Grimwood, and Loyd S. Swenson, Jr. This illustrated history by a trio of experts is the definitive reference on the Apollo spacecraft and lunar modules. It traces the vehicles' design, development, and operation in space. More than 100 photographs and illustrations. 576pp. 6 3/4 x 9 1/4. 0-486-46756-2

EXPLORING THE MOON THROUGH BINOCULARS AND SMALL TELESCOPES, Ernest H. Cherrington, Jr. Informative, profusely illustrated guide to locating and identifying craters, rills, seas, mountains, other lunar features. Newly revised and updated with special section of new photos. Over 100 photos and diagrams. 240pp. 8 1/4 x 11. 0-486-24491-1

WHERE NO MAN HAS GONE BEFORE: A History of NASA's Apollo Lunar Expeditions, William David Compton. Introduction by Paul Dickson. This official NASA history traces behind-the-scenes conflicts and cooperation between scientists and engineers. The first half concerns preparations for the Moon landings, and the second half documents the flights that followed Apollo 11. 1989 edition. 432pp. 7 x 10.
0-486-47888-2

APOLLO EXPEDITIONS TO THE MOON: The NASA History, Edited by Edgar M. Cortright. Official NASA publication marks the 40th anniversary of the first lunar landing and features essays by project participants recalling engineering and administrative challenges. Accessible, jargon-free accounts, highlighted by numerous illustrations. 336pp. 8 3/8 x 10 7/8. 0-486-47175-6

ON MARS: Exploration of the Red Planet, 1958-1978--The NASA History, Edward Clinton Ezell and Linda Neuman Ezell. NASA's official history chronicles the start of our explorations of our planetary neighbor. It recounts cooperation among government, industry, and academia, and it features dozens of photos from Viking cameras. 560pp. 6 3/4 x 9 1/4. 0-486-46757-0

ARISTARCHUS OF SAMOS: The Ancient Copernicus, Sir Thomas Heath. Heath's history of astronomy ranges from Homer and Hesiod to Aristarchus and includes quotes from numerous thinkers, compilers, and scholasticists from Thales and Anaximander through Pythagoras, Plato, Aristotle, and Heraclides. 34 figures. 448pp. 5 3/8 x 8 1/2.
0-486-43886-4

AN INTRODUCTION TO CELESTIAL MECHANICS, Forest Ray Moulton. Classic text still unsurpassed in presentation of fundamental principles. Covers rectilinear motion, central forces, problems of two and three bodies, much more. Includes over 200 problems, some with answers. 437pp. 5 3/8 x 8 1/2. 0-486-64687-4

BEYOND THE ATMOSPHERE: Early Years of Space Science, Homer E. Newell. This exciting survey is the work of a top NASA administrator who chronicles technological advances, the relationship of space science to general science, and the space program's social, political, and economic contexts. 528pp. 6 3/4 x 9 1/4.
0-486-47464-X

STAR LORE: Myths, Legends, and Facts, William Tyler Olcott. Captivating retellings of the origins and histories of ancient star groups include Pegasus, Ursa Major, Pleiades, signs of the zodiac, and other constellations. "Classic." – *Sky & Telescope*. 58 illustrations. 544pp. 5 3/8 x 8 1/2. 0-486-43581-4

A COMPLETE MANUAL OF AMATEUR ASTRONOMY: Tools and Techniques for Astronomical Observations, P. Clay Sherrod with Thomas L. Koed. Concise, highly readable book discusses the selection, set-up, and maintenance of a telescope; amateur studies of the sun; lunar topography and occultations; and more. 124 figures. 26 halftones. 37 tables. 335pp. 6 1/2 x 9 1/4. 0-486-42820-6

Browse over 9,000 books at www.doverpublications.com

CATALOG OF DOVER BOOKS

Chemistry

MOLECULAR COLLISION THEORY, M. S. Child. This high-level monograph offers an analytical treatment of classical scattering by a central force, quantum scattering by a central force, elastic scattering phase shifts, and semi-classical elastic scattering. 1974 edition. 310pp. 5 3/8 x 8 1/2. 0-486-69437-2

HANDBOOK OF COMPUTATIONAL QUANTUM CHEMISTRY, David B. Cook. This comprehensive text provides upper-level undergraduates and graduate students with an accessible introduction to the implementation of quantum ideas in molecular modeling, exploring practical applications alongside theoretical explanations. 1998 edition. 832pp. 5 3/8 x 8 1/2. 0-486-44307-8

RADIOACTIVE SUBSTANCES, Marie Curie. The celebrated scientist's thesis, which directly preceded her 1903 Nobel Prize, discusses establishing atomic character of radioactivity; extraction from pitchblende of polonium and radium; isolation of pure radium chloride; more. 96pp. 5 3/8 x 8 1/2. 0-486-42550-9

CHEMICAL MAGIC, Leonard A. Ford. Classic guide provides intriguing entertainment while elucidating sound scientific principles, with more than 100 unusual stunts: cold fire, dust explosions, a nylon rope trick, a disappearing beaker, much more. 128pp. 5 3/8 x 8 1/2. 0-486-67628-5

ALCHEMY, E. J. Holmyard. Classic study by noted authority covers 2,000 years of alchemical history: religious, mystical overtones; apparatus; signs, symbols, and secret terms; advent of scientific method, much more. Illustrated. 320pp. 5 3/8 x 8 1/2.
0-486-26298-7

CHEMICAL KINETICS AND REACTION DYNAMICS, Paul L. Houston. This text teaches the principles underlying modern chemical kinetics in a clear, direct fashion, using several examples to enhance basic understanding. Solutions to selected problems. 2001 edition. 352pp. 8 3/8 x 11. 0-486-45334-0

PROBLEMS AND SOLUTIONS IN QUANTUM CHEMISTRY AND PHYSICS, Charles S. Johnson and Lee G. Pedersen. Unusually varied problems, with detailed solutions, cover of quantum mechanics, wave mechanics, angular momentum, molecular spectroscopy, scattering theory, more. 280 problems, plus 139 supplementary exercises. 430pp. 6 1/2 x 9 1/4. 0-486-65236-X

ELEMENTS OF CHEMISTRY, Antoine Lavoisier. Monumental classic by the founder of modern chemistry features first explicit statement of law of conservation of matter in chemical change, and more. Facsimile reprint of original (1790) Kerr translation. 539pp. 5 3/8 x 8 1/2. 0-486-64624-6

MAGNETISM AND TRANSITION METAL COMPLEXES, F. E. Mabbs and D. J. Machin. A detailed view of the calculation methods involved in the magnetic properties of transition metal complexes, this volume offers sufficient background for original work in the field. 1973 edition. 240pp. 5 3/8 x 8 1/2. 0-486-46284-6

GENERAL CHEMISTRY, Linus Pauling. Revised third edition of classic first-year text by Nobel laureate. Atomic and molecular structure, quantum mechanics, statistical mechanics, thermodynamics correlated with descriptive chemistry. Problems. 992pp. 5 3/8 x 8 1/2. 0-486-65622-5

ELECTROLYTE SOLUTIONS: Second Revised Edition, R. A. Robinson and R. H. Stokes. Classic text deals primarily with measurement, interpretation of conductance, chemical potential, and diffusion in electrolyte solutions. Detailed theoretical interpretations, plus extensive tables of thermodynamic and transport properties. 1970 edition. 590pp. 5 3/8 x 8 1/2. 0-486-42225-9

Browse over 9,000 books at www.doverpublications.com

CATALOG OF DOVER BOOKS

Engineering

FUNDAMENTALS OF ASTRODYNAMICS, Roger R. Bate, Donald D. Mueller, and Jerry E. White. Teaching text developed by U.S. Air Force Academy develops the basic two-body and n-body equations of motion; orbit determination; classical orbital elements, coordinate transformations; differential correction; more. 1971 edition. 455pp. 5 3/8 x 8 1/2. 0-486-60061-0

INTRODUCTION TO CONTINUUM MECHANICS FOR ENGINEERS: Revised Edition, Ray M. Bowen. This self-contained text introduces classical continuum models within a modern framework. Its numerous exercises illustrate the governing principles, linearizations, and other approximations that constitute classical continuum models. 2007 edition. 320pp. 6 1/8 x 9 1/4. 0-486-47460-7

ENGINEERING MECHANICS FOR STRUCTURES, Louis L. Bucciarelli. This text explores the mechanics of solids and statics as well as the strength of materials and elasticity theory. Its many design exercises encourage creative initiative and systems thinking. 2009 edition. 320pp. 6 1/8 x 9 1/4. 0-486-46855-0

FEEDBACK CONTROL THEORY, John C. Doyle, Bruce A. Francis and Allen R. Tannenbaum. This excellent introduction to feedback control system design offers a theoretical approach that captures the essential issues and can be applied to a wide range of practical problems. 1992 edition. 224pp. 6 1/2 x 9 1/4. 0-486-46933-6

THE FORCES OF MATTER, Michael Faraday. These lectures by a famous inventor offer an easy-to-understand introduction to the interactions of the universe's physical forces. Six essays explore gravitation, cohesion, chemical affinity, heat, magnetism, and electricity. 1993 edition. 96pp. 5 3/8 x 8 1/2. 0-486-47482-8

DYNAMICS, Lawrence E. Goodman and William H. Warner. Beginning engineering text introduces calculus of vectors, particle motion, dynamics of particle systems and plane rigid bodies, technical applications in plane motions, and more. Exercises and answers in every chapter. 619pp. 5 3/8 x 8 1/2. 0-486-42006-X

ADAPTIVE FILTERING PREDICTION AND CONTROL, Graham C. Goodwin and Kwai Sang Sin. This unified survey focuses on linear discrete-time systems and explores natural extensions to nonlinear systems. It emphasizes discrete-time systems, summarizing theoretical and practical aspects of a large class of adaptive algorithms. 1984 edition. 560pp. 6 1/2 x 9 1/4. 0-486-46932-8

INDUCTANCE CALCULATIONS, Frederick W. Grover. This authoritative reference enables the design of virtually every type of inductor. It features a single simple formula for each type of inductor, together with tables containing essential numerical factors. 1946 edition. 304pp. 5 3/8 x 8 1/2. 0-486-47440-2

THERMODYNAMICS: Foundations and Applications, Elias P. Gyftopoulos and Gian Paolo Beretta. Designed by two MIT professors, this authoritative text discusses basic concepts and applications in detail, emphasizing generality, definitions, and logical consistency. More than 300 solved problems cover realistic energy systems and processes. 800pp. 6 1/8 x 9 1/4. 0-486-43932-1

THE FINITE ELEMENT METHOD: Linear Static and Dynamic Finite Element Analysis, Thomas J. R. Hughes. Text for students without in-depth mathematical training, this text includes a comprehensive presentation and analysis of algorithms of time-dependent phenomena plus beam, plate, and shell theories. Solution guide available upon request. 672pp. 6 1/2 x 9 1/4. 0-486-41181-8

Browse over 9,000 books at www.doverpublications.com

CATALOG OF DOVER BOOKS

HELICOPTER THEORY, Wayne Johnson. Monumental engineering text covers vertical flight, forward flight, performance, mathematics of rotating systems, rotary wing dynamics and aerodynamics, aeroelasticity, stability and control, stall, noise, and more. 189 illustrations. 1980 edition. 1089pp. 5 5/8 x 8 1/4. 0-486-68230-7

MATHEMATICAL HANDBOOK FOR SCIENTISTS AND ENGINEERS: Definitions, Theorems, and Formulas for Reference and Review, Granino A. Korn and Theresa M. Korn. Convenient access to information from every area of mathematics: Fourier transforms, Z transforms, linear and nonlinear programming, calculus of variations, random-process theory, special functions, combinatorial analysis, game theory, much more. 1152pp. 5 3/8 x 8 1/2. 0-486-41147-8

A HEAT TRANSFER TEXTBOOK: Fourth Edition, John H. Lienhard V and John H. Lienhard IV. This introduction to heat and mass transfer for engineering students features worked examples and end-of-chapter exercises. Worked examples and end-of-chapter exercises appear throughout the book, along with well-drawn, illuminating figures. 768pp. 7 x 9 1/4. 0-486-47931-5

BASIC ELECTRICITY, U.S. Bureau of Naval Personnel. Originally a training course; best nontechnical coverage. Topics include batteries, circuits, conductors, AC and DC, inductance and capacitance, generators, motors, transformers, amplifiers, etc. Many questions with answers. 349 illustrations. 1969 edition. 448pp. 6 1/2 x 9 1/4.
0-486-20973-3

BASIC ELECTRONICS, U.S. Bureau of Naval Personnel. Clear, well-illustrated introduction to electronic equipment covers numerous essential topics: electron tubes, semiconductors, electronic power supplies, tuned circuits, amplifiers, receivers, ranging and navigation systems, computers, antennas, more. 560 illustrations. 567pp. 6 1/2 x 9 1/4. 0-486-21076-6

BASIC WING AND AIRFOIL THEORY, Alan Pope. This self-contained treatment by a pioneer in the study of wind effects covers flow functions, airfoil construction and pressure distribution, finite and monoplane wings, and many other subjects. 1951 edition. 320pp. 5 3/8 x 8 1/2. 0-486-47188-8

SYNTHETIC FUELS, Ronald F. Probstein and R. Edwin Hicks. This unified presentation examines the methods and processes for converting coal, oil, shale, tar sands, and various forms of biomass into liquid, gaseous, and clean solid fuels. 1982 edition. 512pp. 6 1/8 x 9 1/4. 0-486-44977-7

THEORY OF ELASTIC STABILITY, Stephen P. Timoshenko and James M. Gere. Written by world-renowned authorities on mechanics, this classic ranges from theoretical explanations of 2- and 3-D stress and strain to practical applications such as torsion, bending, and thermal stress. 1961 edition. 560pp. 5 3/8 x 8 1/2. 0-486-47207-8

PRINCIPLES OF DIGITAL COMMUNICATION AND CODING, Andrew J. Viterbi and Jim K. Omura. This classic by two digital communications experts is geared toward students of communications theory and to designers of channels, links, terminals, modems, or networks used to transmit and receive digital messages. 1979 edition. 576pp. 6 1/8 x 9 1/4. 0-486-46901-8

LINEAR SYSTEM THEORY: The State Space Approach, Lotfi A. Zadeh and Charles A. Desoer. Written by two pioneers in the field, this exploration of the state space approach focuses on problems of stability and control, plus connections between this approach and classical techniques. 1963 edition. 656pp. 6 1/8 x 9 1/4.
0-486-46663-9

Browse over 9,000 books at www.doverpublications.com

CATALOG OF DOVER BOOKS

Mathematics-Bestsellers

HANDBOOK OF MATHEMATICAL FUNCTIONS: with Formulas, Graphs, and Mathematical Tables, Edited by Milton Abramowitz and Irene A. Stegun. A classic resource for working with special functions, standard trig, and exponential logarithmic definitions and extensions, it features 29 sets of tables, some to as high as 20 places. 1046pp. 8 x 10 1/2. 0-486-61272-4

ABSTRACT AND CONCRETE CATEGORIES: The Joy of Cats, Jiri Adamek, Horst Herrlich, and George E. Strecker. This up-to-date introductory treatment employs category theory to explore the theory of structures. Its unique approach stresses concrete categories and presents a systematic view of factorization structures. Numerous examples. 1990 edition, updated 2004. 528pp. 6 1/8 x 9 1/4. 0-486-46934-4

MATHEMATICS: Its Content, Methods and Meaning, A. D. Aleksandrov, A. N. Kolmogorov, and M. A. Lavrent'ev. Major survey offers comprehensive, coherent discussions of analytic geometry, algebra, differential equations, calculus of variations, functions of a complex variable, prime numbers, linear and non-Euclidean geometry, topology, functional analysis, more. 1963 edition. 1120pp. 5 3/8 x 8 1/2. 0-486-40916-3

INTRODUCTION TO VECTORS AND TENSORS: Second Edition--Two Volumes Bound as One, Ray M. Bowen and C.-C. Wang. Convenient single-volume compilation of two texts offers both introduction and in-depth survey. Geared toward engineering and science students rather than mathematicians, it focuses on physics and engineering applications. 1976 edition. 560pp. 6 1/2 x 9 1/4. 0-486-46914-X

AN INTRODUCTION TO ORTHOGONAL POLYNOMIALS, Theodore S. Chihara. Concise introduction covers general elementary theory, including the representation theorem and distribution functions, continued fractions and chain sequences, the recurrence formula, special functions, and some specific systems. 1978 edition. 272pp. 5 3/8 x 8 1/2.
0-486-47929-3

ADVANCED MATHEMATICS FOR ENGINEERS AND SCIENTISTS, Paul DuChateau. This primary text and supplemental reference focuses on linear algebra, calculus, and ordinary differential equations. Additional topics include partial differential equations and approximation methods. Includes solved problems. 1992 edition. 400pp. 7 1/2 x 9 1/4. 0-486-47930-7

PARTIAL DIFFERENTIAL EQUATIONS FOR SCIENTISTS AND ENGINEERS, Stanley J. Farlow. Practical text shows how to formulate and solve partial differential equations. Coverage of diffusion-type problems, hyperbolic-type problems, elliptic-type problems, numerical and approximate methods. Solution guide available upon request. 1982 edition. 414pp. 6 1/8 x 9 1/4. 0-486-67620-X

VARIATIONAL PRINCIPLES AND FREE-BOUNDARY PROBLEMS, Avner Friedman. Advanced graduate-level text examines variational methods in partial differential equations and illustrates their applications to free-boundary problems. Features detailed statements of standard theory of elliptic and parabolic operators. 1982 edition. 720pp. 6 1/8 x 9 1/4. 0-486-47853-X

LINEAR ANALYSIS AND REPRESENTATION THEORY, Steven A. Gaal. Unified treatment covers topics from the theory of operators and operator algebras on Hilbert spaces; integration and representation theory for topological groups; and the theory of Lie algebras, Lie groups, and transform groups. 1973 edition. 704pp. 6 1/8 x 9 1/4.
0-486-47851-3

Browse over 9,000 books at www.doverpublications.com

CATALOG OF DOVER BOOKS

A SURVEY OF INDUSTRIAL MATHEMATICS, Charles R. MacCluer. Students learn how to solve problems they'll encounter in their professional lives with this concise single-volume treatment. It employs MATLAB and other strategies to explore typical industrial problems. 2000 edition. 384pp. 5 3/8 x 8 1/2. 0-486-47702-9

NUMBER SYSTEMS AND THE FOUNDATIONS OF ANALYSIS, Elliott Mendelson. Geared toward undergraduate and beginning graduate students, this study explores natural numbers, integers, rational numbers, real numbers, and complex numbers. Numerous exercises and appendixes supplement the text. 1973 edition. 368pp. 5 3/8 x 8 1/2. 0-486-45792-3

A FIRST LOOK AT NUMERICAL FUNCTIONAL ANALYSIS, W. W. Sawyer. Text by renowned educator shows how problems in numerical analysis lead to concepts of functional analysis. Topics include Banach and Hilbert spaces, contraction mappings, convergence, differentiation and integration, and Euclidean space. 1978 edition. 208pp. 5 3/8 x 8 1/2. 0-486-47882-3

FRACTALS, CHAOS, POWER LAWS: Minutes from an Infinite Paradise, Manfred Schroeder. A fascinating exploration of the connections between chaos theory, physics, biology, and mathematics, this book abounds in award-winning computer graphics, optical illusions, and games that clarify memorable insights into self-similarity. 1992 edition. 448pp. 6 1/8 x 9 1/4. 0-486-47204-3

SET THEORY AND THE CONTINUUM PROBLEM, Raymond M. Smullyan and Melvin Fitting. A lucid, elegant, and complete survey of set theory, this three-part treatment explores axiomatic set theory, the consistency of the continuum hypothesis, and forcing and independence results. 1996 edition. 336pp. 6 x 9. 0-486-47484-4

DYNAMICAL SYSTEMS, Shlomo Sternberg. A pioneer in the field of dynamical systems discusses one-dimensional dynamics, differential equations, random walks, iterated function systems, symbolic dynamics, and Markov chains. Supplementary materials include PowerPoint slides and MATLAB exercises. 2010 edition. 272pp. 6 1/8 x 9 1/4. 0-486-47705-3

ORDINARY DIFFERENTIAL EQUATIONS, Morris Tenenbaum and Harry Pollard. Skillfully organized introductory text examines origin of differential equations, then defines basic terms and outlines general solution of a differential equation. Explores integrating factors; dilution and accretion problems; Laplace Transforms; Newton's Interpolation Formulas, more. 818pp. 5 3/8 x 8 1/2. 0-486-64940-7

MATROID THEORY, D. J. A. Welsh. Text by a noted expert describes standard examples and investigation results, using elementary proofs to develop basic matroid properties before advancing to a more sophisticated treatment. Includes numerous exercises. 1976 edition. 448pp. 5 3/8 x 8 1/2. 0-486-47439-9

THE CONCEPT OF A RIEMANN SURFACE, Hermann Weyl. This classic on the general history of functions combines function theory and geometry, forming the basis of the modern approach to analysis, geometry, and topology. 1955 edition. 208pp. 5 3/8 x 8 1/2. 0-486-47004-0

THE LAPLACE TRANSFORM, David Vernon Widder. This volume focuses on the Laplace and Stieltjes transforms, offering a highly theoretical treatment. Topics include fundamental formulas, the moment problem, monotonic functions, and Tauberian theorems. 1941 edition. 416pp. 5 3/8 x 8 1/2. 0-486-47755-X

Browse over 9,000 books at www.doverpublications.com

THE KEEPER

USA TODAY BESTSELLING AUTHOR
BELLA MATTHEWS

THE KEEPER

A KROYDON HILLS LEGACY NOVEL

PLAYING TO WIN
BOOK ONE

BELLA MATTHEWS

Copyright © 2023

Bella Matthews

All rights reserved. No part of this publication may be reproduced or transmitted by any means, electronic, mechanical, photocopying, recording or otherwise, without the prior permission of the publisher, except in the case of brief quotation embodied in the critical reviews and certain other noncommercial uses permitted by copyright law.

Resemblance to actual persons, things, living or dead, locales or events is entirely coincidental. The author acknowledges the trademark status and trademark owners of various products referenced in this work of fiction, which have been used without permission. The publication/use of these trademarks is not authorized, associated with, or sponsored by the trademark owners.

This book contains mature themes and is only suitable for 18+ readers.

Editor: Dena Mastrogiovanni, Red Pen Editing

Cover Designer: Sarah Sentz, Enchanting Romance Designs

Photographer: Michelle Lancaster, Lane Photography, @Lanefotograf

Model: Andy Murray

Interior Formatting: Brianna Cooper

SENSITIVE CONTENT

This book contains sensitive content that could be triggering.
Please see my website for a full list.

WWW.AUTHORBELLAMATTHEWS.COM

I've been lucky enough to have my best friend for almost 20 years, but it wasn't until I was in my 40's that I found my tribes.

This book is dedicated to all of you out there still hoping to find your tribes as much as it's dedicated to the ladies that have given me mine. My Captain America girls & my Cohansey Moms. The ladies I text with the funny reels I know they will laugh at with me. The ones I smile with and cry with. The ones who hold me up when I can't stand by myself. The women who celebrate my accomplishments more loudly than I ever could. I love you all and am so grateful to have you in my life.

"Don't settle for the one you can live with, wait for the one you can't live without."

— UNKNOWN

CAST OF CHARACTERS

The Kingston Family

- **Ashlyn & Brandon Dixon**
  - Madeline Kingston - 23
  - Raven Dixon - 7

- **Max & Daphne Kingston**
  - Serena Kingston - 16

- **Scarlet Cade St. James**
  - Brynlee St. James - 22
  - Killian St. James - 20
  - Olivia St. James - 18

- **Becket & Juliette Kingston**
  - Easton Hayes - 27
  - Kenzie Hayes - 21
  - Blaise Kingston - 11

- **Sawyer & Wren Kingston**
  - Knox Kingston - 15
  - Crew Kingston - 12

- **Hudson & Maddie Kingston**
  - Teagan Kingston - 16
  - Aurora Kingston - 13
  - Brooklyn Kingston - 8

- **Amelia & Sam Beneventi**
  - Maddox Beneventi - 21

- Caitlin Beneventi - 18
  - Roman Beneventi - 16
  - Lucky Beneventi - 14

- **Lenny & Bash Beneventi**
  - Maverick Beneventi - 19
  - Ryker Beneventi - 17

- **Jace & India Kingston**
  - Cohen Kingston - 14
  - Saylor Kingston - 10
  - Atlas Kingston - 7
  - Asher Kingston - 7

The Kings Of Kroydon Hills Family

- **Declan & Annabelle Sinclair**
  - Everly Sinclair - 22
  - Grace Sinclair - 22
  - Nixon Sinclair - 21
  - Leo Sinclair - 20
  - Hendrix Sinclair - 17

- **Brady & Nattie Ryan**
  - Noah Ryan - 19
  - Lilah Ryan - 19
  - Dillan Ryan - 16
  - Asher Ryan - 10

- **Aiden & Sabrina Murphy**
  - Jameson Murphy -19
  - Finn Murphy - 16

- **Bash & Lenny Beneventi**
  - Maverick Beneventi - 19
  - Ryker Beneventi - 17

- **Cooper & Carys Sinclair**
  - Lincoln Sinclair - 12
  - Lochlan Sinclair - 12
  - Lexie Sinclair - 12

- **Coach Joe & Catherine Sinclair**
  - Callen Sinclair - 22

For family trees, please visit my website www.authorbellamatthews.com

PROLOGUE

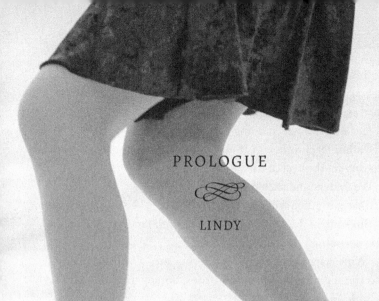

LINDY

Nineteen Years Old

"Hey, princess," Easton's smooth voice whispers as I answer his FaceTime in the middle of the night.

"Hey, hockey boy," I rasp back and clear the sleep from my eyes. "What time is it in Vegas?"

It's too late to be doing time-zone math.

Or maybe it's too early. I squint to see the clock app. *Definitely too early.*

"It's a little after midnight here. Were you sleeping?"

I grab my glasses and sit up so I can see his face. Easton always FaceTimes or texts. He never calls. There's no in-between for him. There never has been. "E, are you drunk?"

He runs his hand through his sandy-brown hair. Hair that looks like it's already been yanked on one too many times. *Eww.* Please don't let a naked woman be in bed next to him. Because I've gotten those calls before, and they are not my favorite. "Easton . . ." I push when he doesn't answer me. "What's going on? Are you alone? Are you okay?"

"I fucked up, Lindy." With haunted eyes, he drops his head back against his pillow and groans. "It wasn't supposed to be like this."

He and I have been doing this for years.

Calling each other in the middle of the night when our demons get the best of us.

We understand each other.

Shared trauma will do that to a person.

But tonight, he's talking in riddles even I'm having a hard time decoding.

"What happened, E? You're scaring me," I whisper softly into the night, as my stomach drops, anticipating the worst possible answers.

"I couldn't save you," he breathes out and shuts down.

"But you *did* save me, Easton. I'm alive because of you." I pull my knees up to my chest and wrap an arm protectively around myself. I never talk about this. Not with anyone except him. "You saved us both." Four years ago, a stalker held Easton and me at gunpoint. In an effort to get to my mother, he killed my bodyguard, and if it hadn't been for Easton and my stepfather, Brandon, he would have killed the rest of us too.

A chill runs down my spine, and I try to shake it off before Easton closes his eyes. "In my dreams, I couldn't save you."

"In my dreams, you always do," I tell him with brutal honesty because *honesty* is the only thing we've ever been able to offer each other. "Are you going to be able to sleep, E?"

"Stay on the phone with me, okay? I need to hear you breathe. I need to know you're safe."

I lie back down and tug my comforter up, then prop my phone on the pillow. "Sleep, E."

This isn't the first time I've gotten this call in the middle of the night.

It won't be the last either.

The Philly Press
KROYDON KRONICLES

SOCIALITES IN SIN CITY

Looks like our favorite Kroydon Hills socialites have jet setted off to sin city for baby Kingston's birthday celebration. Let's see if what happens in Vegas stays in Vegas, or if this reporter can bring the dirt back to Philly.
#KroydonKronicles

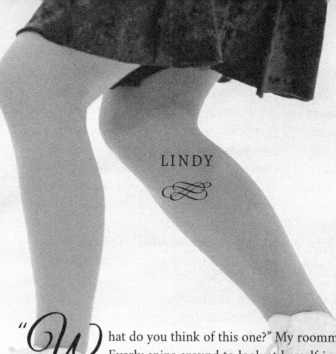

LINDY

"What do you think of this one?" My roommate Everly spins around to look at herself in the mirror. Her blonde curls bounce around her shoulders like a shampoo commercial.

"Hurry up," her twin sister, Gracie, yells from the hall of the penthouse suite we've booked for the weekend.

Everly rolls her eyes and plumps her perfect boobs, and not for the first time, I wish I had my friend's confidence . . . *and figure*. Give me a pair of figure skates and a perfectly smooth rink and I've got confidence in spades. In all the other facets of my life, *fake it till you make it* is a little more my speed. So I go for the easy out and toss a pillow at Everly.

"Who are you worried about impressing? This is a girls' weekend. *Remember?*" With a shake of my head, I lean back on the bed, and maybe I peek down at my own less than C-cup chest. Who am I kidding? I'm barely a B most days. "Remind me again why I'm sharing a room with you?"

She adds a delicate gold chain around her neck and gives me one of her *cheerleadery* smiles. "Because you love me, and I'm going to play wingman for you this weekend. Now get up and get dressed because there are at least two gorgeous men meeting us at the pool that are of legal age and not related to either of us." With another quick glance in the mirror, my Instagram-perfect bestie slides her feet into heeled wedges and tosses a sexy pink sarong my way. "Now rip off that stupid old t-shirt and put this over your bikini. Easton's never going to realize you're an adult with actual boobs and an ass if you dress the same way you did when you were fourteen."

"Evie," I gasp and look around, checking to see if anyone else heard what she said. "I told you that in confidence."

"You told me that in a drunken fucking cryfest, and Kenzie and Brynlee are probably studying in the other room," she snaps back at me, referring to our other two roommates—Kenzie, Easton's little sister, and Brynlee. Technically, I guess Bryn is *Kenzie's* cousin. And considering Kenzie's legal guardian happens to be my big brother, that may actually make me her aunt. I'm the youngest in a huge, obnoxious family. Most days, I wouldn't trade them for the world. Even if my closest sibling in age is fifteen years older than me.

But while I've always looked at Kenzie as part of the family, I've never looked at her brother that way.

Even as a kid, Easton was always handsome. What I had back then was an innocent crush on my best friend's brother. But the feelings that have developed over the years since *that night* are anything but innocent. They're also not something I'll ever admit. At least not sober, apparently.

"Besides, Easton won't be the only man out there today, and you need to pop that cherry at some point. Vegas is bound to have plenty of men willing to do the work for you," she winks dramatically and smiles again.

I wrap a hand around Everly's big mouth, then yank it away just as fast when she licks it. "*Gross.* You know you have a problem with volume control . . ." I toss my tee onto the hotel bed and tie the sarong around my waist, then fix my boobs. "And who exactly are the two men you're expecting?"

Gracie walks in, the complete opposite of her identical twin sister. Where Everly is a wild bombshell, Grace has a quiet, understated beauty about her. She's the calm to Everly's storm and is always the most sensible one of us. "Oh come on, Lindy. Evie's always wanted Easton's friend, Pace. Do you have any idea how excited she was when she heard he was going to be here this weekend?"

I shrug and refuse to admit I got butterflies when I thought about seeing Easton this afternoon. He and I talk at least once a week and have for years. But we rarely get to see each other. Between my training and his long-ass season, not to mention the Olympics and the two of us living on opposite sides of the country, I hardly ever see him in person.

"Time to let loose, Lindy," Everly giggles and pushes me forward. "Why else did I push for us to come to Vegas for your birthday?"

I guess I could use one of those crazy weekends. Not that it's going to be one of *those* weekends. "I figured it was for the private jet that comes with me."

Everly shoves me. "Not gonna lie. It's awfully convenient to have an heiress for a best friend. But I was just excited your little break from skating meant we could finally plan something for your birthday. Thirty-six hours is better than nothing, and you know me . . . Go big or go home." She smiles like the Cheshire freaking cat. "I plan on going home knowing exactly how *big* Pace is."

I look between my two friends and laugh. Their father has been the franchise quarterback for the professional football team my family owns for over twenty years. Their grandfather is the head coach, and his youngest son, Callen, who happens to be the same age as the twins, was just drafted. None of us come from average families. Unfortunately, I'm the only one who also comes with an around-the-clock bodyguard. "Is Charles waiting outside?"

Brynlee pops her head into the room. "Of course he is. Now let's go. I want to get a cabana at the pool while we still can. That stupid Kroydon Kronicles column in the *Philly Press* has already reported we're in Vegas. Let's not give them anything else to report on."

"I already booked it," I tell her as I grab my bag and toss my Kindle inside. "We've got two for the day. Plenty of room for low-key." I look over at Everly. "At least for those of us that want it."

We walk into the living room of the suite and find Kenzie studying on the couch. Everly snags the big book out of her hands. "You promised us one weekend, girl genius."

"Guys . . . classes are kicking my ass." Kenzie looks up, clearly frustrated. "Exams are next week."

Everly pulls her up to her feet, then hands Kenzie her oversized bag before eventually relenting and letting her add the textbook to the bag. "You know, they have these things called e-books now, Kenz. You should try it."

Kenzie adds her highlighter, then checks to make sure she's got everything else she needs. "Listen . . . you all graduated and started your lives. I've got three years left of med school before I even start a residency." She spins on Everly with a finger pointing her way. "And don't say it. I know I chose this. I've got a test on Monday, and it's only my first one of the week. I have to do well. So I don't want to hear a single word about it when I'm studying at the pool and you're all getting drunk." She looks around at all of us this time, then grabs her vibrating phone, and her face drops. "Ugh . . . Looks like Easton is bailing on the pool, but he said he'll meet us at the club tonight."

"Did he say why?" Everly asks.

Kenzie shakes her head.

Brynlee takes Kenzie's phone and drops it in her bag. "Oh well, his loss. Callen and Maddox are already down there. Let's go."

"I swear if Callen goes home and tells Dad anything about this weekend, I'm going to string up his balls by the laces of his cleats," Everly grumbles.

"Dramatic much?" Brynlee plants her hands on her hips and glares because our girls will always defend Callen.

Forget Maddox. Bryn would throw him to the wolves.

But Callen . . . He's a whole other story.

"Come on," Gracie groans before she pops a big straw hat on her head and grabs the key card from the table. "Can we go now? I'm hungry."

"Oh, sweetie, we're drinking breakfast. Your options are mimosas or bloody marys. But hey, olives are a veggie, now let's go." Everly opens the door with a flourish and pats my bodyguard, Charles, on the chest. "You're going to keep your distance today, right, Chuck?"

Charles's eyes find mine, and I can already tell it's going to be a long weekend. "You know the rules, Miss Sinclair."

Everly leans into my side and whispers, "You know we're going to have to ditch him at some point, right?"

A small smile tugs at my twitching lips.

Maybe it will be one of *those* weekends after all.

Easton

"Why did we even bother taking the meeting if you knew you were going to decline the trade offer? Max Kingston has been trying to get your ass to play for the Revolution for years." Pace has been a broken record about this for the past two hours. "You ever planning on accepting the trade? On going home? Could you even imagine the way the fans would lose their shit over the prodigal son returning?"

"I *am* home, asshole. Vegas has been my home for a decade. Kroydon Hills . . . Well, Kroydon Hills is the place I visit. That's it." I look around to make sure no one's paying attention and shove my best friend forward. "Seriously? Prodigal son? Who the fuck says that kind of shit?"

"There's a reason I'm your agent, E-man. I can spin shit into gold." He's not lying either. Pace was a good college hockey player, but he knew it wasn't going any further, so he went into the family business. Now he's one of the most sought-after sports agents in the country. He's second only to his older brother, which has always pissed him off. "Now relax. It's time to see your girl."

I stop dead in my tracks, and Pace almost runs into me. "Stop with the *my girl* shit."

"Dude. What crawled up your ass tonight? I've been calling *that girl* your girl for a fucking decade. She's the goddamn Kingston princess, for fuck's sake, and she looks like—"

I cut him off with an icy glare. "Watch it."

He throws his hands up in front of himself. "Whatever you gotta tell yourself to get through the night, man. You can keep lying to yourself if you want to, but I'm your best friend. You can't lie to me."

"There's nothing there. End of story. Now let's get this shit over with." I shake my head and move around some rowdy asshole, not giving a shit that he's sloshing beer all over himself while a prostitute grinds her ass against his dick. Dumb fuck probably doesn't realize she's pay-to-play. Maybe he just doesn't care. I remember those days. I spent a fuck-ton of nights drinking to forget, and there's no better place to do that than the city of sin.

Spent half those nights on the phone with *her* too.

Somewhere along the way, she became my girl.

Not that I said it.

Not that she knew it.

But she fucking felt like it.

"She's not my fucking girl," I mumble again as a blonde bombshell catches my eye on the other side of the purple rope designating the VIP area. She's dancing like she knows she's drawing every single man and woman's attention her way, but that's not why I notice her. No. It's the woman she's dancing with. The tiny wisp of a woman, barely five foot two with long dirty-blonde curls and eyes the color of a stormy Bermuda ocean. The one who has her hands raised in the air and her perfect ass up against Everly, shaking her hips in a way that makes my cock hard. The one who's all woman now.

"Yeah, buddy." Pace smacks my chest. "Keep telling yourself she's not your girl." Pace smiles devilishly at a passing VIP waitress and orders two Macallen 18s. "Now don't mind me while I go make myself a sandwich."

"What the fuck—" He moves before I finish my question, and I watch for a moment as the fucker does exactly what he said he'd do and slides between Everly Sinclair and Lindy. *My* fucking Lindy.

The waitress comes back and hands me two glasses, then steps closer and bats her long, fake lashes. "Is there anything else I can get for you, Mr. Hayes?"

Of course, she recognizes me.

Being a professional hockey player used to have it's perks.

Not anymore.

Now, it's just exhausting.

Now, I just want to play good hockey and be left the fuck alone.

"Umm, you can get your skanky ass off my brother." Kenzie moves in front of me, blocking the waitress, then turns and throws her arms around my neck. "Easton," she squeals. "I missed you so much."

I wrap an arm around my little sister and squeeze. "Hey, Kenz. Missed you too."

She pulls back and smacks my chest. "Why did you bail earlier? I thought I was getting a whole day with you. I wouldn't have flown all the way out here just for tonight."

"Sorry. Meeting with team management." I don't bother adding anything about the trade offer. Max Kingston, the GM and part owner of the Philadelphia Revolution, who also happens to be Lindy's oldest brother, might tell her at some point, but I don't want to see the disappointment on her face if she hears it from me.

The Kingstons like to take care of their own, and luckily, they claimed Kenzie and me when our mom died more than a decade ago and our cousin, Juliette, took us in as our legal guardian. She married Becket Kingston a month later, and the family has claimed us ever since.

Most of them, at least.

I remind myself, again, that Lindy's off-limits.

She always has been. She always will be.

I look over again at Lindy and see red.

She's laughing at something Pace is whispering in her ear.

Fucker.

She might feel like mine, but that girl deserves more than me.

The Philly Press

KROYDON KRONICLES

SUN SOAKED DAY DRINKING

Day drinking and soaking in the sun seemed to keep our girls tame today. But rest assured, this reporter has it on good authority they're hitting up a club tonight. And we all know how these girls like to party. Stay tuned peeps.
#KroydonKronicles

LINDY

"*D*on't people usually celebrate *twenty-one* in Vegas, baby Kingston?" Pace's hot breath tickles my ear.

"Lay off the *baby Kingston* thing, Pace." I laugh and push him back.

"You're two years late, aren't you?" He spins me around to face him, and I catch Everly's wink over his shoulder.

"Is that my friend asking or my former agent?" I tease.

"Fuck former," he argues and grabs my hand. "You're my favorite gold medalist, Lindy. If you're really sure you don't want to compete anymore, there are other options. Did you get the offer I sent you from ESPN? They want you, baby—"

I glare, and he stops before using that stupid nickname.

"I'm retired, Pace. You and Andrew have to accept that at some point." I lean in and kiss his cheek, then glance at my best friend and whisper, "Now have some fun tonight, but don't break any hearts, okay?" Giving him a little shove toward Everly, I step back.

"I'm not the heartbreaker here, Kingston."

Brynlee grabs my hand and tugs me behind her over to the table where Maddox and Callen both sit, legs spread wide, taking up as much room as possible. Two girls sit on either side of Callen, and another sits on Maddox's lap.

"Seriously . . ." Brynlee grabs the bottle of champagne chilling in a bucket and takes a big drink before she refills both our glasses. Champagne sloshes over the top as she points at Callen and Maddox. "You two look like you're one tiger away from a really bad night," she tells the guys, and I choke on my champagne.

Gracie moves next to Bryn and lifts her glass for a refill, then looks at the guys and scrunches her face. "They both look like they're one night away from having to get a shot of penicillin."

I giggle. "I don't know. Maddox could rock the whole Mike Tyson tattoo thing," I tease. "But no losing Doug. Got it?"

Bryn snickers and sips more champagne. "Pretty sure you'd be Doug, Lindy."

I think that makes me the boring one. "I could be worse things." I shrug, and Charles, who's standing a few feet behind us, catches my eye. His thick arms are crossed over an even thicker chest. Not exactly blending in here. More like screaming *bodyguard*. He looks at me with a warning, and I turn away.

Whatever.

Tonight is for the girls, and I'm going to ignore my babysitter and have as much fun as I can squeeze into the remaining hours.

I lift my glass and tap it against Brynlee and Gracie's. "Cheers, girls."

"Cheers." They tap back, and I can't help the smile spreading across my face when Kenzie moves next to the table with Easton by her side. He looks . . . better than good. He looks incredible. Easton's a big guy, at least six foot four with broad shoulders and a thick chest that tapers down to a lean waist. He makes Charles look small.

He's mouthwateringly delicious, and he knows it. And when he smiles at me, I melt a little inside.

Without hesitation, I move around Kenzie, and step into Easton.

He slowly wraps me in his arms, and I bury my face in his chest.

I never feel as safe as I do when I'm with him.

"Hey, princess. Happy birthday," he whispers, and a chill skates down my spine while I breathe him in. This man . . .

I don't say anything.

I can't.

I just enjoy the moment for what it is, knowing it'll be over all too soon. And like a dream you wake up from before you get to the good parts, Easton pulls away and bro-hugs Maddox and Callen as Pace and Everly join us. The waitress stops by and drops off another round of shots, then hands a bottle of expensive tequila to Maddox and tells him it's on the house before smiling and walking away. Lucky for her, she moves quickly because the girl on Maddox's lap looks like she's got claws and isn't afraid to use them.

Kenzie and Brynlee pass out shots before Everly raises hers high in the air. "Happy birthday, Lindy."

"Happy birthday, Lindy," is echoed by our friends, and my eyes momentarily find Easton's before I make my wish and swallow my shot.

Gracie takes my hand in hers and guides me out onto the dance floor as "In Da Club" by 50 Cent plays over the speakers, and the DJ announces, "This song is for the birthday girl, Lindy."

Gracie holds my hand high above her head as the girls join us, and we get lost in the music. We move together with ease, laughing and smiling. Hands wandering. Feeling eyes on us and not caring because Charles is here. So is Maddox. And he's a badass in his own right, even if we don't ever talk about it. But it's the other set of eyes on me that burn straight down to my core.

I add an extra sway and bounce that might be just for Easton, but fuck it. It's my birthday, and I don't care. Maybe that's the shots talking. *Oh well.*

I try not to torture myself with thoughts of him like that, but tonight, it doesn't feel like torture. No. Tonight, it's fun. *Tonight,* I'm in control. And I like it.

The five of us belt out the lyrics of each new song, lost in the electric energy buzzing around us, only stopping each time our server brings a new round of what feels like a never-ending round of shots to us. With each new drink, we become louder and more brazen.

Hands slide. Asses shake. Bodies grind.

Maddox still has *claws-out girl* sitting on his lap. Callen is dancing with two girls next to us. And Pace and Everly seem lost in each other on the dance floor.

Guess our girl didn't need a wing woman.

After a few songs and a few more drinks, my skin grows damp, and my hair hangs heavy against the back of my neck.

I lift it off my shoulders, trying to cool down, but it's no use when the shots have done their job and the warm alcohol courses thick through my veins, giving me nerves of steel. I glance back at the table and get a rush of adrenaline when I find Easton staring back.

His eyes are glued to me. Heavy and hungry.

Only a moment later, they change, and it's not hunger I see.

It's something else. *Someone* else.

Hands slide to my hips.

Big hands. But not the ones I want.

I tear my eyes away from Easton and look over my shoulder at the man who just slid in behind me.

"I like the way you move." He tucks me into him, and for a hot second, I think about grabbing one of the girls and telling him I'm not interested. That's the smart thing to do. The responsible thing. But I don't do that. I look up at him and smile instead.

He's gorgeous.

Not really my type.

A little too preppy, but hot in a country club way.

Expensive clothes, expensive cologne, and I think *maybe* better eyebrows than I have.

He looks like he spends more time getting ready than I ever have, and that's not the kind of guy I go for. But for some reason, I decide to ignore the warning signs that this guy is not for me and lean back against Mr. Tall, Dark, and Handsome.

Maybe Everly's right.

Maybe it's time to let loose.

Easton

Lindy might be dancing with someone else, but she's watching me while she does it. And what the hell does it say about me that I can feel myself getting ready to snap?

With every shake of her ass, the strings of the invisible line I've refused to cross for fucking years are pulled tighter. Each time she lets this douche touch her, another thread frays.

"What the fuck is she doing?" I growl, and Maddox

Beneventi raises his head from his girl's neck. He watches me throw back the rest of my whiskey and tracks what I'm looking at. This douche is grinding against Lindy's ass, but she doesn't seem to mind. Her glassy eyes are closed as his hands roam over her body. Hands that don't belong there. Hands I want to rip from his arms, leaving fucking bloody stumps.

"Chuck's here. Lindy's fine," Maddox tells me as I slam my glass back down.

He scoops his girl off his lap and hands her a hundred. "How about you go get us another round, babe?"

Her eyes light up like she just hit the fucking jackpot. "Okay," she squeals, and I'm glad she's with him and not me.

"Calling her *babe* because you don't know her name, madman?"

"Might want to slow down there, E-man. You're give-a-fuck is showing."

"The hell you talkin' about, man?" I've known Maddox a long damn time. As long as I've known Lindy and the Kingstons. But he's younger than me, and it's not like we hang out together, braiding each other's hair. He's always been a cocky fucker. But when your dad runs the Philly mob, I guess that's what happens.

He pours the last of the Macallen into two glasses and hands me one, then swallows his. "I'm talking about the way you watch Lindy, asshole. I'm talking about the way you've *been* watching her since she was a kid. Since we were all kids, you included. She might be blind, but I'm not. And judging by the way you look like you're about ten seconds away from killing the dude she's dancing with, I'm thinking you're not blind either."

I look over at her again and grit my fucking teeth as her head falls back on this douche's shoulder.

Man, it's easier to ignore this fucking thing between us

when she's across the country, living her life, and I'm here, living mine.

Fuck this.

I slam my glass down on the table and push back from my chair.

"Be sure, man," Maddox warns. "She's the baby, and the whole damn family still sees her that way."

"Guess it's a good goddamned thing I'm not part of your family then."

The bastard sits back in the booth and smiles.

Fuck him and fuck this.

I move onto the dance floor and grab Lindy's hand. "We need to talk."

The douche's eyes grow wide when he sees me. Recognition lighting them up.

Guess he's a Vegas Vipers fan.

I'd bet my signing bonus he knows exactly who I am.

"You're..."

"I am, buddy. Now how about you give us a minute?" I pull Lindy toward me and watch the way her stormy eyes darken as her hands run up my arms.

"Easton," she breathes out but doesn't push me away. "That was rude."

I bend my knees and toss her over my shoulder. "Then I guess I'm sorry about *this*."

"Easton," she calls out, laughing. Damn, I love that sound.

"What's going on, brother?" Pace asks with Everly glued to his side.

"We're getting the hell out of here, man," I tell him and start walking, knowing her whole crew, including my sister, is following behind.

LINDY

I'm not sure what wakes me up first . . . the throbbing in my head or the obnoxiously loud alarm I don't remember setting on my phone. I yank the pillow over my face to drown it out, but it's no use.

Wait . . . I think that's a ring tone.

Who the hell is calling this early?

I swing my hand out, trying to silence the phone and smack my wrist against the corner of the nightstand instead.

Ow. That hurt. Not enough to stop the pounding in my head, but enough.

I'm pretty sure you're not supposed to be able to feel your pulse behind your eyes.

This is not normal.

I lie with my eyes closed, trying to piece together why the hell I feel this way, but last night is fuzzy. Almost as fuzzy as my mouth. *Eww.* The last thing I remember was . . . *Shots.* Dancing. *More shots.* Then what?

"Relax, princess. I got you."

That voice . . . *Oh my God.* I've heard that voice more times than I can count, but it's never sounded quite *that* good. Deep and gravelly and so fucking sexy that heat pools in places it has no business pooling at the moment.

A big, warm, deliciously callused palm wraps around my waist and presses flat against my stomach. My very bare stomach. Butterflies take flight, and every nerve-ending in my entire body stands alert. And that's before I'm pulled back against an incredibly firm chest, and the man that chest belongs to groans.

A man who shouldn't be in my bed.

Why *is* he in my bed?

Wait . . . is this my bed?

"Unless you don't want to sleep," Easton murmurs as he buries his face in my hair as that question hangs in the air.

Fuck me.

Wait. *No.*

This has got to be a dream. I'm on a girls' trip in Vegas.

I'm rooming with Everly.

Nowhere in my plan was I supposed to end up in a bed with anyone this weekend.

Especially. *Not.* Him.

No . . . I press the pillow down against my eyes.

This can't be happening.

It's a dream. *You're still dreaming.*

Hips press against my ass, and any doubt that I might actually still be dreaming quickly vanishes because in my dreams, Easton Hayes doesn't feel this good. Of course, my dreams usually end before I get the chance to enjoy his ridiculously large erection pressing firmly against my ass.

I shift a little, and Easton's hands grip my hips. "Lindy," he warns.

This. Cannot. Be. Happening.

"Yeah, princess, it is."

Huh?

Who's he answering?

"You, baby. Now stop thinking so loud and go back to sleep." Easton pulls the pillow off my face and tucks *it* and his arm under my head, positioning me so I'm snuggled between the crook of his neck and his bicep.

Just where I always wanted to be, only I have no idea how the hell I got here.

How many times have I wondered what this would feel like? And now that I know, how am I ever going to live without it again? Easton's mouth presses against my neck, and a small moan slips past my lips.

Stupid, traitorous lips.

This isn't right.

Maybe nothing happened.

Maybe he just fell asleep next to me.

Or maybe I finally indulged in the one thing I've always wanted to do but never had the lady balls to grab for myself.

Okay, time to be a big girl. Roll the fuck over and face the music.

I take a hot fucking second to cringe at the poorest excuse for a pep talk I've ever given myself, and I've given myself plenty. I'm a goddamn gold medalist. I can do pep talks. They just usually happen on the ice or in the locker room. Occasionally in a car. Once while lying in the wet grass when I fell running and had to convince myself to get the hell back up and finish the run. But never in my wildest dreams did I think I'd be giving myself one in bed.

Stalling done, I try to carefully roll over without exposing any of my bits in the process, and two things happen at once. First, I say a quick thank-you to the one-night-stand gods because as I roll over, my panties go straight up my ass in the most uncomfortable way possible. Sleeping in a thong is not fun. But I'm pretty sure if I had sex with Easton last night, my panties would have been incinerated in the process. I'm hoping this means I didn't finally give up my virginity when I was sloppy drunk to the man I've been half in love with since before I started shaving my legs.

The second I look up, any thoughts about how my thong is permanently wedged up my ass like dental floss or about how drunk I must have been last night evaporate into thin air. Because Easton is looking at me with the sexiest smile I've ever seen. *Wow*. That smile promises wicked things. "Mornin', princess."

He presses his lips to my forehead, and I'm pretty sure I melt into a puddle of goo, right here on the thousand-count Egyptian cotton sheets on the massive hotel bed. My headache forgotten, I bring a shaky hand up to his neck and dig my fingers into the back of his hair.

For a single second, I let myself lie here, safe in his arms before panic sets in.

Because it always sets in.

I pull back, yanking the blanket up around my chest to cover myself while inching back against the headboard. "What the hell, E?"

Easton runs his hand up my thigh, and damn it, there go those goosebumps again, followed by a literal knee-jerk reaction when he tickles me.

As in, *maybe* I kick him a little.

And maybe he kinda, sorta falls off the bed.

Because really, how many more ways could this morning be more humiliating?

Easton falls to the floor, tangled up in the blanket with a thud, and I peek over at him. "What the fuck, Lindy?"

I can't believe this is happening.

I close my eyes as embarrassment washes over me, followed by freezing cold waves of panic. With a deep breath, I hide my face in my hands. Only, when I yank my hand back, I stare in horror at the big, fat, perfect brilliant-cut diamond sitting on my ring finger, right next to a matching band.

A wedding band.

My mouth opens and shuts a few times as I try to find words. Then I look from the beautiful diamond and platinum band to the mouthwatering man now standing at the foot of the bed, shirtless and in a pair of navy-blue boxer briefs. Every inch of his golden chest is on beautiful display. Muscles stretch under taut skin. Veins bulge. It's a sight I would love to savor if it weren't for the shock I'm pretty sure I'm going into. Because there's a plain black band on his left ring finger too.

"My eyes are up here, princess."

I snap my head up to his stupid grin and throw a pillow at his face as I climb up to my knees. "Wanna tell me why I have a wedding ring on my finger, Easton Hayes?"

"Pretty sure because you're my wife, Madeline Hayes."

"I'm sorry. WHAT?" I shriek at Easton as I stand up and attempt to secure the sheet around myself, while hysteria bubbles underneath my skin. "For a second, I thought you said I was your wife. But that couldn't be right. I mean, that's crazy." I fight to get the stupid fucking sheet knotted so I can move without my boobs popping free but can't seem to manage since my hands won't stop shaking. "I can't be your wife. I'm not even your girlfriend." When I still can't get the damn sheet tied, I grab a white t-shirt off the floor and take a step toward Easton. "How exactly could I possibly be your wife?"

Easton takes the shirt out of my hands and pulls it down over my head like I'm a freaking child, and I manage to slide my arms through it without flashing him. *My husband.* "What the fuck, E?"

His eyes soften as I drop the sheet to the floor and step out of it. The shirt comes to mid thigh, covering all the important bits, and I feel slightly better for a second until he reaches out and cups my face. "What's the last thing you remember, Lindy?"

I close my eyes—trying to ignore how good it feels to be held like this—and try to focus on last night, but that makes my head hurt ten times worse. "Everything gets a little fuzzy after the shots."

Oh, lord. So many shots. "There was dancing."

"Yeah, baby. There was dancing. A lot of dancing," he murmurs as his thumb rubs along my cheekbone.

I lean into it, and then my eyes fly open. "You threw me over your shoulder like I was a bag of dirty laundry," I exclaim, and a sexy laugh rumbles in his throat.

"You throw a lot of bags of laundry over your shoulder, princess?" He bends his knees, bringing his forehead to rest against mine when I don't laugh at his stupid joke, fighting to hold back the tears threatening to pool in my eyes instead. "Come on, Lindy. Try to remember what happened after that. I need you to remember the rest of the night."

I shake my head and immediately regret the motion as my head threatens to explode again. "How were you sober, E? You had as many shots as I did."

"I wasn't completely sober. But I wasn't blackout drunk. And I didn't think you were either. At least, not then. When we all piled onto the party bus afterward, and you and Everly started chugging champagne straight from the bottle, I thought maybe you wouldn't be feeling too great today." He tilts my face up to his, and I'm shocked by the hurt I see there. "But I wasn't expecting you not to remember anything."

"East—" I'm cut off by a banging on the hotel room door.

"Madeline Kingston, kiss your husband goodbye and get your ass moving. The jet leaves in an hour." When I don't answer her right away, too busy being stuck on the fact she just told me to kiss my husband, she bangs again. "I've been calling you all morning. Now answer the damn phone or open the damn door."

Guess it was Everly who woke me up earlier.

That's one question solved.

Only about a million more to go.

I step back from Easton, cross the room, and crack the door open. "Give me a minute, okay?"

She stuffs her hands through the crack in the door and shoves clothes at me. "Hurry up and say goodbye to lover boy. We're waiting on you, and I need to know what to tell my mom."

"Your. What?" I whisper, and my breath is ripped from my body. "Your mom? Your mom knows? Does *my* mom know?"

Please, dear sweet baby Jesus in the manger.

Please, *please*, please, don't let my mom know.

"Everly," I yell and open the door. Only, instead of Everly being there, the hall is empty.

Son of a—

My eyes fly to Easton. "It's Vegas. We can get this annulled, and nobody will ever have to know, right?"

If they don't already.

My mom's gonna kill me.

My sisters . . . Good grief. My brothers are going to lose their minds.

"We've got to get this annulled. Quick," I add on for good measure as I pull up the jeans Everly handed me.

"No."

"I'm sorry. What?" I ask as I stare in disbelief. "What the hell do you mean *no*?"

Easton crosses the room in two strides and sinks his hand into my hair, pulling me closer.

I drop whatever else I was holding as I lean back against

the door behind me and lay my palms over his chest. "Easton—"

"Stop talking, princess." He brushes his mouth over mine, and a million sparks light up my body for the very first time. I sigh, and Easton's tongue pushes into my mouth. Firm and deliciously demanding. Making me momentarily forget about this morning. About any war I was about to wage. I ignore the fear and anxiety bubbling under the surface and just feel him. Feel. Us. Until suddenly I can't feel him anymore.

Because Easton pulls away, leaving my body cold and my heart racing. "Like I said, princess. Like it or not, we're married, and we're going to stay that way until you remember last night. Once you can tell me you remember marrying me, if you still want to annul this, I will. But for now, go pack your bags." He smacks my ass and turns me toward the door. "See you soon, wife."

Easton

The door closes with a soft snick behind Lindy, and I have the overwhelming urge to open it back up, just so I can slam it shut.

How the fuck does she not remember last night?

I bared my fucking soul.

She fucking said—

My phone rings, cutting off my thoughts, and I yank it off the table. "What?"

"Good morning to you too, asshole."

"I'm not in the mood for your shit today, Pace." I hit

speaker and toss the phone on the bed so I can find my clothes.

"Aww. The honeymoon sex a disappointment? It gets better, buddy. You'll last longer next time."

"There was no sex last night, dick." This motherfucker. "Lindy woke up this morning, didn't remember any of it, and asked for an annullment."

"Ohhh . . . Burn. That had to be a blow to your big, fat ego, huh? How you gonna fix it? Not sure it can really be fixed. But you're gonna try, right?"

"You gonna take a breath, man?" I drop down on the bed and tie my damn boots, then rest my elbows on my knees. "She's getting on a jet in an hour. How am I supposed to fix anything?"

"You want to give her the annulment?"

"Fuck no," I growl.

Pace scoffs like the shithead he is. "Weren't you trying to convince me yesterday she *wasn't* your girl?"

Trying to convince myself is more like it, but I keep that to myself. "I was wrong, and you know it. You've always known it. And you've never missed a chance to point it out."

"I know. I just wanted to hear you say it. Can you hold off on the annulment until you can talk to her? Or better yet, see her? When's the next time the Vipers play the Revolution? At least then, you'll be in the same city."

"Pace, you're a fucking genius. I could kiss you, man." I jump up and grab my keys and the phone.

"Dude. I've always been a genius. 'Bout time you fucking noticed. But I'm gonna need you to tell me what the fuck you're talking about this time."

Always the smartass. "I need you to make a call." I smile as I walk out of the hotel room. "I need you to accept the trade."

The Philly Press

KROYDON KRONICLES

MADELINE KINGSTON MARRIES EASTON HAYES

Buckle your seatbelts and put those tray tables in an upright position, peeps, because I'm about to take you on a ride. Breaking news this morning: Madeline Kingston is now Mrs. Madeline Hayes! If a certain socialite's social media is to be believed, baby Kingston married Vegas Vipers' hotshot goalie, Easton Hayes. This isn't the first time these two have made headlines together, so you might want to get ready for a bumpy ride.
#KroydonKronicles

LINDY

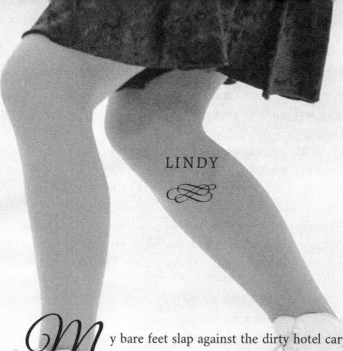

My bare feet slap against the dirty hotel carpet as I stomp out of the elevator on my floor. And yes, I know how gross that sounds. And that I'm probably not the first person to do this disgusting walk of shame. *Ick*.

That thought hits about the same time I manage to walk face-first into Maddox's chest. His hands grip my shoulders as he laughs. "Hold up, trouble. Where's the fire?"

"You're a tool, Maddox." He grins, and I pinch his nipple. We're *very* mature for our age. Technically, I may be his aunt, but he's always been more like an annoying brother than anything else. He's two years younger than me but likes to act like he's ten years older.

The door to our suite opens behind him, and I cringe.

Great. More witnesses to my humiliation.

Brynlee pops her head out of our door and scrunches her nose like she just smelled a skunk. Or more accurately, just got a look at my morning-after face, which is probably even worse than I'm imagining. Because seriously, that would be about right. "Come on, Lindy. If you move fast, you've got time to shower before we have to leave for the airport." Her nose scrunches again. "And use some extra body wash. The booze is wafting from your pores."

Great.

I look like a hot mess, and apparently, I smell like a bar.

Freaking fabulous.

My phone vibrates in my hand, and I silence it without even glancing down.

This day is off to a stellar start.

Maddox ignores Bryn and shoulders his leather weekender bag with his stupid, cocky smirk. "See you at the airport, *trouble*."

"Giant tool," I mumble under my breath and push into the suite, where my best friends all stand, waiting for me with varying shades of *what the fuck* flashing across their faces.

Huh. Is this what an intervention feels like?

I cannot deal with this. Not now. Not with the excitement vibrating through Everly or the concern coming off Bryn and Grace. And worse, not when the disappointment's clear as day on Kenzie's face.

Nope. Can't. Deal.

I put my hand up before anyone opens their mouth. "I'm going to take a shower. A long one. A hot one. *Really* hot. Scalding hot. Any chance someone ordered Starbucks for the road?"

"Lindy, stop." Gracie moves around the girls. "You can't run away from us."

The room spins as the pressure builds behind my eyes. "I just need a minute." I grab her hand, desperate for her to hear more than just the words. "Just a minute to breathe. I'm not running away."

Everly tilts her head with a wicked grin in her eyes, and I glare. "Zip it, Evie." She probably knows me better than anyone, so she knows I shut down. And okay, maybe I tend to ignore or run away from my problems. But I really just need five fucking minutes to shut down. Alone.

I turn back and feel like I've been sucker-punched by the hurt on Kenzie's face.

How many people am I going to hurt today? "Listen, I love you. And I know we need to talk, but right now, I need a shower. I need some fucking coffee, and I need to stop feeling my pulse behind my eyeballs." My voice raises with each new word until I'm full-blown yelling. "We've got an entire country's worth of a flight home, where you can spend hours telling me how stupid I am. Trust me. I'm already disgusted with myself. But I beg you, please. For the love of all that's holy. *Please* give me a few fucking minutes before you start the lecture." I take a few steps before turning back. "And not a word of this to a single soul in the family or I will disown you all."

Everly clears her throat, but I thrust a finger in the air, cutting her off. "Not. One. Word."

I don't bother waiting for an answer or looking back again as I move into the bathroom and lock the door for good measure. My friends don't really do *boundaries*.

I try to forget the expressions on their faces as I slide down the shower wall and wrap my arms around my knees, but it's not that easy because when I don't see their faces, Easton's is everywhere. Hot tears mix with the spray of scalding hot water while I sob silently.

Since I was a little girl, I've wanted to marry Easton Hayes.

Wanted the white dress and the long aisle.

Wanted to wake up next to him and know he was mine. Really mine.

I'm pretty sure I doodled *Mrs. Madeline Hayes* a time or ten in a notebook after I met him the very first time. He was so handsome and so broody. It's hard to forget just how broody he was back then. I knew no boys would ever compare to him. And that was before.

Before I even knew what an incredible man he'd become.

Or how much he'd mean to me.

Never in a million years did I fantasize about waking up next to him with a ring on my finger and having absolutely no memory of how it got there.

I wished for this to be the year my life finally changed.

The year I got what *I* wanted.

Stupid birthday wish.

The girls gave me a wide berth when I got out of the shower.

They left me alone as I threw my clothes back into my bag, handed me a coffee, and stayed quiet as we drove to the airport. What do they say about small miracles?

It's not until we're all on the plane and Everly sits down next to me that she decides I've had enough time to sulk and lifts my sunglasses off my head. "We need to talk."

I close my eyes and lean back against the leather seat. "I know. And I'm sorry I yelled before. It's not your fault. It's not any of your faults. I'm the idiot. It's just . . ." I try to put into words the insane emotions warring inside me. The hurt. The devastation. The anger. At myself. At Easton. "I don't know what I'm gonna do."

My phone rings again, and I silence it for the millionth time this morning and toss it in my purse.

"Well." Everly reaches inside my bag and pulls it back out. "I'd say you're going to have to answer this at some point. But there's something you need to know first."

I crack open my eyes. "What else could I possibly need to know? Did I have a threesome last night too?"

"Jesus Christ, trouble," Maddox groans way too loud. "I don't need that shit burned into my brain."

"You picture Lindy having sex a lot, madman?" Everly taunts. "*Kinky*. I like it."

Callen takes a swig from a flask and cracks an arrogant smile. "Don't knock it till you try it."

"Try what? Picturing Lindy having sex?" Brynlee asks, shocked.

"Eww." Grace's cheeks pink as she smacks Callen's shoulder. "Just. Eww."

Callen winks at me. "What? I'm not related to her."

"Oh my God." Everly covers her face as she cracks up.

"I didn't have a threesome," I moan. "It was a bad joke." Because that's what this whole day has been. *A bad joke*. "We're going to get it fixed, and if I'm lucky, we're going to do it before the whole world finds out."

"Easton isn't something that needs to be fixed, Lindy," Kenzie clips back, protecting her brother.

"Lindy . . ." Brynlee pushes when my phone keeps ringing.

"Oh my God." I give up and yank the stupid thing out of Everly's hand. "Hello?"

"Madeline Kingston. What the hell were you thinking?"

And the hits just keep coming.

A collective groan echoes around the cabin.

Gracie reaches across the aisle and hands me her iPad, and I gasp and completely miss whatever else is being said. Because on the screen is Everly's Instagram page.

With a post from last night.

Easton's holding my face in his hands, and I'm holding his hands in mine.

And that look in my eyes. *Wow*. We look so happy. So . . . in love.

And then there are the rings on my finger. They're on full display for the world to see.

Everly's caption reads *Congratulations to the new Mr. and Mrs. Easton Hayes!*

I look at the girls and cringe. "Hi, Mom."

LINDY

As Charles pulls up to our building, Brynlee squeezes my hand. "You sure you want to go right over to your mom's? Maybe you should take some time to think first."

We landed at the private airport outside of Kroydon Hills about an hour ago, and the knot that had been slowly growing in my stomach during our flight doubled in size. Everly basically live streamed my wedding, and it's been picked up by a handful of news outlets already, including the *Philly Press* and ESPN.

My mom is going to kill me.

I bite down on the inside of my cheek until I taste the metallic tang of blood. "No. But I'm not sure I have a choice. If I don't go, Brandon and she will just show up here."

"True. Then she might bring reinforcements," Everly adds as she and the girls climb out of the SUV.

"Yup. Like my mom," Bryn snickers. "Suck it up, buttercup. Better to get it over with. and hopefully avoid the rest of the family." She follows the others out of the SUV, and I drop my head against the seat.

"Traitor," I murmur as they greet our doorman.

We used to share a house closer to the main campus of Kroydon University, but once my family's company purchased this small, seven-unit building, I was conveniently offered the top floor. They combined two units and converted them into a five-bedroom condo. I guess you could say my family has control issues, and this was a way to maintain control.

Maddox and Callen sharing one of the condos one floor below us was another way. There are four other apartments on the two floors beneath them, and a coffee shop and gym on the first floor. It's a great place to live, but I'm not naive enough to think this place doesn't come with strings.

My family has always been overprotective.

That kicked into overdrive after everything with Mom's stalker.

Like I said, control issues.

Charles looks back at me through the rearview. "You ready to go, Madeline?"

"That seems to be the million-dollar question, doesn't it?" Or at least one of many.

"I think your mother will have a few other questions." He pulls away, and my stomach drops.

I fight the urge to laugh because I'm afraid once I start, I won't be able to stop. "Any chance you can take the long way home?"

"I'll see what I can do."

Unfortunately for me, we hit every greenlight on our way to Mom and Brandon's. Our town isn't that big, so there's only three lights to go through, but still. Would it have been too much to ask the universe for just one of these suckers to be red?

As we pull into the private lakeside estates where half my family lives, I smile at all the holiday decorations lighting up the gorgeous houses. It's barely December, and already, there's a hint of snow on the ground and ice on the lake. It soothes my soul. This is my favorite time of year. My mother's house is covered in white twinkly lights and green wreaths with red-velvet bows hanging from the windows. Small white candles light each one, giving off a warmth I know will change the second I step inside.

As soon as I wrap my hand around the front doorknob, my little sister, Raven, cracks it open with a finger over her mouth. "Shh. Mommy is in the kitchen with Aunt Lenny and Aunt Scarlet, and they're talking about you."

My bulldog, Myrtle, runs down the stairs and over to us, excited to see me. Miss overdramatic acts like I've been gone for two months instead of two days.

I squat down and squish them both to me. "How do they sound?" I ask quietly.

Her big brown eyes look up at me. "They sound loud," she whispers, and I almost laugh. *Almost.*

"Madeline . . ." My oldest sister, Scarlet, also known as Brynlee's mom, walks out into the hall and stops a few feet away. Her crimson lips press tightly together, and I prepare myself for the hit she's about to throw my way. "Is your *husband* with you?"

And there it is.

"Hey, Raven, how about you go find Dad, okay?" I run my hand over her silky black hair and nudge her down the hall. "And take Myrtle with you, please." When she turns toward the stairs, I stand back up. "Just me tonight, Scar."

She gently shakes her head and wraps an arm around my shoulder. "Well, at least you did one thing right."

She walks me into the kitchen like it's my walk down the green mile, and my mom and Lenny both stand waiting for me at the end. The only one missing is—

My brother's wife, Juliette, comes out of the basement

with two bottles of wine in her hands. "Found them." She looks from Mom to me and hands me the bottles. "How about you pour us some wine, and then you can fill us in on how exactly you became my pseudo-daughter-in-law."

I take both bottles from her and glance at my mother, who hasn't said a word to me. "Have you talked to Easton?" My voice shakes, betraying my nerves.

Easton and Kenzie moved in with Juliette and my brother Becket when their mom died years ago. E and Juliette have always had a special relationship, which according to Kenzie, has only gotten stronger over the years.

"I've talked to his voice mail, if that counts. But he hasn't called me back. So I thought it would be better to come straight to the source. Right now, that's you, kiddo. And I figured someone needed to be here to stop your mother and Scarlet from murdering you." Juliette hands me the bottle opener and looks at my mom.

"I thought that was *my* job," Lenny offers, but I ignore her and face the music, also known as my mother.

"Hi, Mom."

The mask of indifference she's wearing slips and her exasperated glare zeroes in on me. "What were you thinking?" Her tone is sharp enough to cut glass.

This is going to be so much worse than I thought.

My entire life, I've been the good girl.

The one who always did the right thing and always did what I was told.

I spent a lifetime building a level of trust with Brandon and her.

And in one night, I destroyed that.

"Madeline . . . I . . . I just don't." She rips one of the bottles out of my hands and turns her back on me as she opens it herself. Once she fills her glass and swallows it in three gulps, she turns back slowly. "I'm trying to stay calm, but I'm not sure I can," she tells me, slightly more in control than she was a moment ago. "I don't understand what you were thinking. Are you acting out? Are you on drugs? Is this because you gave up skating and now you're floundering, trying to figure out what you want to do with your life?"

Ouch. That hurts.

"I need you to explain this to me because I'm having a really hard time trying to understand what in the ever-loving hell you were thinking." She gasps and covers her mouth. "You're not pregnant, are you?"

"No, I'm definitely not pregnant," I answer, mortified.

"Ashlyn," Lenny whispers, and Mom's fiery eyes fly to hers.

Len takes the other bottle from my hands as I stand there, frozen in place, certain my mother has never been this disappointed in me before. "Sit down, Lindy." She pulls out one of the counter stools and pushes me into it, then pours me a glass of wine.

"That's it, Len. Reward her with more alcohol. Because I'm sure she didn't have enough last night when she married Easton in a dirty chapel in Las Vegas," Scarlet taunts and yanks the bottle out of Lenny's hands, then sets her sights on me. "If you'd at least warned me, I could have gotten in front of this with the press. Haven't we taught you anything?"

"The press?" I squeak. Then I think about the *Philly Press* and ESPN articles I saw earlier. Son of a bitch. If there's already two, there's bound to be more.

"Yes, Madeline," Mom snaps like one of those dragons from *Game of Thrones* before it opens its mouth and decimates an entire city with one fiery breath. "The *press*. You are one of the wealthiest heiresses in the entire country. An Olympic gold medalist. You have how many million social-media followers? Did you think the press wouldn't take notice when you married the boy who saved your life? The one who happens to be one of the top goalies in the entire hockey league and whose social-media presence rivals yours?"

Lenny sips her wine. "Maybe if Everly hadn't posted a picture."

"Maybe if that stupid Kroydon Kronicles column wasn't obsessed with the whole group of you . . ." Juliette adds.

"Or maybe if you had behaved like an adult instead of a reckless, irresponsible child." My mother levels me with a hard stare.

"You know what?" I slowly stand, attempting to hide my rapidly shredding confidence. I should tell them to back off. That I'm twenty-three and have never given them a reason not to trust me or my judgment. Remind them that it's Easton, and he'd never do anything to hurt me. But I can't.

Although, I think I needed that last reminder myself.

For a hot second, I think about telling them all to shut up.

But that's not going to fix anything.

My family doesn't know how to shut up.

I'm not even sure it would make me feel better.

Instead, I decide to tuck my tail between my legs and act like their version of the adult they want me to be. "I'm sorry I let you down. I'll talk to Easton, and we'll get this taken care of."

I move to leave the kitchen but stop without turning around when my mom calls out my name. "Where do you think you're going?"

"To kiss Brandon and Raven goodbye and grab my dog. I'm exhausted."

"You're just going to leave?" Mom's tone wavers for the first time tonight, and I almost feel bad for what I've put her through. *Almost*. But I don't. Because this is my life. And not a single person in this room bothered to ask me if I wanted to marry Easton Hayes.

No one asked me if I loved him, or if he loved me.

They've all just assumed I was a drunken idiot.

Which, okay, so maybe I was.

But for my entire life, I've been the good girl. The smart girl. The girl who trained harder, longer, and more often than anyone else. I've been the perfect daughter. Perfect partner. And the perfect Kingston. What I've never been is irresponsible.

I deserved more from them than this tonight. But I'm not going to waste my breath trying to argue that point because everyone in this room still sees a baby instead of a grown woman. And married or not, that's not going to change.

I walk out of the kitchen and find Brandon leaning against the wall at the end of the hall, silently listening. His strong arms are crossed over his chest, but when he sees me, he immediately opens them and pulls me in for a hug. "Hey, shortcake. How are you feeling?"

"Like I just got run over by a stampede of wild horses." I bury my face against him and somehow manage not to cry. "I've never seen her so mad before."

"Your mom's just worried about you. This isn't like you, Lindy. Hell, it's not like Easton either. What's going on?"

I close my eyes and soak in my stepfather's strength. "I'm not actually sure yet."

Brandon kisses the top of my head, then rests his chin there.

My lip trembles while I fight back the tears.

"Guess you better figure that out then, shouldn't you?"

I nod. "Yeah. I guess I should."

The apartment is dark and quiet when Myrtle and I get home later that night. My lazy little bulldog moves surprisingly fast when she runs into our living room, sticks her fat face in her toy basket, pulls out her favorite stuffed dinosaur dressed in a Kings jersey, then settles on her fluffy bed in front of the fireplace, and starts snoring within seconds. She was never a super-active dog, but she's definitely slowed down a bit this year. She and I have been together since I was fifteen, and I may actually love her more than a few members of my family.

Okay, well, maybe just Maddox.

I grab a bottle of water out of the kitchen and make sure the place is locked up before heading to my room. Judging by the lights out, I guess everyone crashed early, which sounds pretty good to me. "Lindy . . ." I stop at Kenzie's door. It's cracked open with a soft glow coming through.

"Hey."

She closes her laptop and makes room for me next to her. "How were the moms?"

"Even Juliette was there." I pull back her blanket and crawl into bed next to her. "They're all so mad at me, Kenz."

"Kinda like you were with Easton earlier?"

"Touché." I link my pinky with hers and lay my head on her pillow. "I'm sorry. I know I put you in a funky spot. I just wish I could remember last night. He remembered everything but refused to tell me any of it. Then told me he wouldn't give me an annulment. I *was* just so mad. I still am. But I'm not sure if I'm mad at him or at myself."

She doesn't say anything, but she doesn't need to. The look on her face is enough.

"I know I didn't handle it well. But Kenz, it's Easton. *Easton*," I plead and hope she understands what I'm saying.

It's easy to close my eyes and go right back to *that night*.

To the way he held me while that psychotic man held a gun to my head.

The way he kept us both safe.

To all the phone calls all the nights since.

"It's *Easton*, Kenz," I plead again. "There's no playbook for this. I don't know what I'm supposed to do."

She shoves a hand under her face and shocks me when she smiles.

"Why are you smiling?"

She rolls her lips together, then smiles again. "Because, Linds. It's like you said. It's Easton. *Easton and you*. Technically, we're sisters now."

"Yeah." I tug the blanket up higher. "I guess we are." I cringe because I always wanted this. But never this way.

"Just try to keep an open mind when you talk to him. If I know my brother, there's more to it. But you need to talk to him to get the answers you want." She rolls over and clicks off the light on the nightstand. "Go to sleep, Linds."

"How am I supposed to sleep when I've got the distinct impression you know more than you're telling me?" I toe off my fuzzy socks under the blanket, then pull my hoodie over my head and toss it to the floor. "I really wish one of you Hayes siblings would fill me in."

"Try calling him tomorrow. Maybe you'll get your answers."

Yeah. Maybe I will.

*T*he next day I try to work up the nerve to call Easton and fail miserably.

In my defense, he doesn't call me either.

So maybe I decide to take the coward's way out.

LINDY

We need to talk.

EASTON

Do you remember marrying me yet?

LINDY

No. That's what we need to talk about.

We shouldn't have gotten married.

EASTON

I disagree.

LINDY

Easton . . . How can you say that? I don't even remember marrying you.

EASTON

That's exactly why I can say that. I know what happened Saturday night, princess. And it wasn't a mistake. If you can tell me you remember it too and still want an annulment, I'll give you an annulment.

LINDY

Why are you being difficult?

EASTON

Because you're worth it.

Gotta go, wife. I've got a game to play.

Because you're worth it.

Damn him.

LINDY

What do you do when you can't shake a funk?

EVERLY

I remember that I have a great ass and things could be worse.

BRYNLEE

This is why we're friends.

GRACIE

You do have a great ass.

EVERLY

You're just saying that because we have the same ass.

KENZIE

You just made me snort Coke out of my nose.

LINDY

WHAT!

KENZIE

The soda. Come on. It's not like Evie said it.

EVERLY

I've never snorted anything up my nose, thank you very much.

LINDY

I'm laughing so hard I can't breathe.

EVERLY

Good. Funk gone. Now on with your day.

The Philly Press

KROYDON KRONICLES

BICOASTAL BABY WATCH

For those of you following the newest super couple, Kroydon Hills' favorite goalie was on fire last night during the Vegas Vipers game against the Colorado Crush. Still waiting to see if this couple is going to be bicoastal. I'd hate to see that jet fuel bill. I don't know, peeps. Should we be on baby watch? Stay tuned and see.
#KroydonKronicles

EASTON

"I've been expecting your call, kid. Thought it would have come a little earlier than this though."

"Listen, old man, I'm pushing thirty. I think it's time you retired the whole *kid* thing," I tell Becket as I wait for my Uber outside the Philadelphia airport. "And it's only been a few days."

"Rumor has it you're old and married, at least according to Juliette and a few hundred gossip sights." Yeah . . . Becks has always had a way with words. "And it's been almost a week."

"Looks that way." Not that my wife has returned a single one of my calls since she texted me the day after she left Vegas. And I've been calling. "And you're arguing semantics. A few days. Almost a week. I say tomato . . ."

My Uber pulls up, and I slide into the back seat and confirm the address with the driver.

"Why are you giving someone my address, Easton?" Becks questions.

"Because I'm coming home, Becket."

"Home? Like *home*, home?" he asks, and okay, yeah, maybe now I do feel like a kid again because this is how Becks used to question me before I moved out.

"Home. Like *Kroydon Hills* home, Becks. Like Max offered me a trade this weekend, and I accepted it, *home*. We had to hammer out a few details, but it was official as of this morning."

Becks sucks in an audible breath and blows it out in a long, low whistle. "Well damn, kid. It took you long enough. Max has been making that offer for a long time. Glad to know you finally took it. And a little pissed he didn't tell me."

"Yeah. I know. I can feel the *I told you so* vibes through the phone. And don't be mad at Max. I asked him to let me break the news." I look out the window at the dusky, snow-covered city and think about all the reasons I wasn't ready to come home until now. Most of them starting and ending with Lindy.

"Maybe I wasn't talking about the trade," he taunts. "You've got the family pretty upset with Lindy and you. You ready to talk about that yet?"

"You ready to listen? Because the messages Jules has been leaving me don't really sound like she's ready to hear me out. She just sounds pissed."

"Listen, Juliette may not be your mom, but she loves you and Kenzie the exact same way she loves Blaise. It doesn't matter that she didn't give birth to you. You're hers. *Ours*. And she's hurt. We didn't exactly expect to find out from social media that Lindy and you got drunk and married in Vegas. It's safe to say some of the family may be a little pissed off."

Becket Kingston came into my life a week after my mother died. I was an angry teenager, mad at the world. He didn't try to change me or fix me or force me into some fucked up box that would fit the Kingston mold. Instead, he spent time getting to know me. Setting boundaries and proverbially knocking me down whenever I stepped over them, which I did, *a lot*. He never raised a hand and rarely raised his voice. *No*. He used his words. He led by action and demanded I follow. That's how Becks works. Probably why he's spent the past decade as a US senator.

He earned every ounce of respect I will forever have for him. Even if I don't tell him enough. So if he wants to be pissed I drunkenly married his baby sister in Vegas, he's earned that right. But it's not going to change anything. Madeline Kingston is my wife. And I don't care what anyone else in this family thinks about that.

"You pissed, Becks?"

"Kid, I may be the only one who isn't. My wife . . . ? Now, *she's* pissed. And according to Jules, your wife is pretty pissed at everyone too. Not sure if that includes you."

Lindy's pretty pissed with me at the moment. At least judging by the lack of communication, it's pretty safe to assume she is. Although it's probably not a great idea to tell Becks that. "Am I better off asking why you're not . . . or why Jules is?" I ask as the driver pulls onto Main Street in Kroydon Hills.

Becks sighs. "I've been married long enough to know better than to speak for Juliette. But for me, let's just say I have faith in the man you are, and *that* man wouldn't marry my sister without loving her. You wouldn't do that to Madeline or to me. You're a good man, Easton. Now, do I wish you hadn't done it drunk in Vegas? Yeah, fuckhead. I've been trying to calm Jules down for twenty-four hours. Newsflash —it's not working. And she's not half as upset as Ashlyn. *But* —and this is a big one—if you tell me right now you want to be married to my baby sister—if you tell me she's it, and you're willing to take on the whole family—I'll back you up 100 percent. I'll fight the family with you. You're not a dumb kid, E, and I'm not oblivious enough to ignore the fact that there's always been something between the two of you."

The driver pulls up to the closed gates in front of Kingston Manor, the ten-thousand square-foot mansion where I spent my last two years of high school living with Jules and Becks. "You think you can buzz me in?"

"So where are Jules and Blaise?" I ask Becks as he hands me a bottle of beer.

"She's bringing Blaise home from basketball practice. They'll be here soon." He slides onto the chair across from me at the kitchen table. "Does your sister know you're home yet?"

"No. Just you and Max. I wanted to talk to Jules before I told anyone else. Unless Max already activated your freaky family phone tree."

Becks smiles and shakes his head, letting me know Max hasn't done that. At least *not yet*. "So what's your plan, kid?"

I run my hand over the condensation forming on the bottle and think about my next move. Which is basically the same thing I've been overthinking for the past twenty-four fucking hours. When I look back at Becks, I want to wipe that shit-eating grin off his face.

"Come on. Tell me you didn't change teams and fly across the country without a plan to win Lindy over? Didn't I teach you better than that?" He looks utterly amused when I don't answer. At least one of us is. "Since you're here *alone*, I'm assuming you need to convince my baby sister to remain your wife."

"Maybe," I mumble, and this fucker laughs at me.

"How drunk were you?" He pushes with an edge to his

tone. "Is that why she's pissed? Or did you do something else to upset her?"

I'm not ready to talk about this yet. Not with Becks. Not with anyone. Not until I talk to Lindy. So instead, I lift my head and look at the closest man I've ever had to a father and give him the only thing I can. "It's always been her, Becks."

"I'm not the one you have to convince."

He's right. But he's a safe place to start.

Becks stares at me for a minute, contemplating something. "Do you have a place to stay tonight?"

"No. I was gonna grab a room at the hotel in town."

"Maybe you should go say hi to your sister first," he challenges. "You know they have a third-floor loft that has a spare bed in it. And if you need something more permanent, we haven't filled one of the condos on the floor below theirs. Maddox and Callen are in the other one, so we haven't been in a rush to fill their neighboring unit."

I push back from my chair and grab my bag from the floor as my plan starts to come together. "You're a fucking genius, Becks." With a quick hug and slap to his back, I make my way to the door before he can stop me. "Tell Jules I'll stop by after I check in with the Revolution tomorrow."

"Coward," he laughs as I open the door.

I'm not a coward.

I'm a man on a mission.

"Hey, Becks . . . Any chance I could borrow a car?"

Lindy

"Kenz . . ." I call out as I turn the TV on and find the show. "We're about to start." Myrtle and I snuggle up on the couch in front of the fire with a bowl of popcorn and a bottle of wine for me and a special cookie for her.

Bryn reaches over and grabs a handful of popcorn, dropping a few on the couch, which Myrtle inhales immediately. "Come on. I'm pressing play."

"I'm coming," Kenzie calls back from the kitchen before she walks in with a plate of nachos. "Did you start?"

"Not yet."

We have a somewhat unhealthy addiction to a teen soap opera about a group of football players at prep school called *The Kings Of Kroydon Hills*. It's based on a book series the twins' Aunt Nattie wrote. The whole town went nuts when it was released on a streaming service last year. The only bad thing was we binged all twelve episodes in one weekend and had to wait an entire year for season two. It finally dropped today.

There's a knock at our door, and Bryn and I look at Kenz. It's her turn to answer since she was the last to sit down. House rules.

"Fine. I've got it. But you better pause it."

"The guys aren't coming near us tonight. They know we were planning on watching this."

I throw a piece of popcorn at Bryn. "Like the guys would knock."

"True."

We both turn when Kenzie squeals. "Easton? What are you doing here?"

Brynlee grabs my hand, flipping the popcorn bowl over onto the floor in the process, as the two of us spin around on the couch and stretch to look down the hall.

You have *got* to be kidding me.

Bryn smacks me and silently mouths, *Oh my God.*

The door slams shut, and Bryn and I are too shocked to turn around and act like we're not freaking out when Kenzie and Easton walk into the room. "Hey, E." Bryn smiles and nudges me, trying to get me to close my mouth. "What are you doing here?"

Easton looks from her to his sister before finally setting those gorgeous hazel eyes on me. "I live here now," he tells us, and his still boyish smile, crooked and handsome, stretches across his face. "I've been traded to the Revolution. I thought my sister and my *wife* might want to know."

"You what?" I practically scream at the same time Kenzie pulls him into a hug.

"Finally. It took you long enough," she tells him.

Bryn elbows my ribs again and mouths, *Oh shit.*

Yeah . . . Oh shit doesn't really cover it.

"Did you tell Jules and Becks yet?" Kenzie asks, and Easton drops a bag I hadn't noticed to the floor.

"Better question," I interrupt.

That damn smile gets even bigger, and the dimple that's always done stupid things to my heart pops deep in his left cheek. "Yes, wife?"

"Could you please stop calling me that?" I demand because holy shit, I really don't want to like the way that sounds coming from him.

"Was that your question?" he taunts.

My blood boils. *Was that my question?* "No, that was not my question, smartass. It was *a* question. One I'd like answered. But no. My question is where are you planning to stay?"

"We have an extra room," Kenzie offers, and for the second time in just a few minutes, my head feels like it spins 360 degrees.

"Kenzie," Bryn cuts her off, but it's too late. The damage is done.

Easton throws his arm around his sister and drops a kiss on the top of her head. "Thanks, Kenz. I'll only need it for a night or two. I'm working on lining a place up."

"You've got to be kidding me." I see red as I squat down and shoo Myrtle away from the spilled popcorn she's devouring and grab the bowl. "Can I see you in the kitchen, please?"

Kenzie's wide eyes fly to mine. "Me?"

"No." I dip my head toward Easton. "Him." I grab his hand to tug him behind me, furious with him and Kenzie. But he apparently didn't get the memo because the big jerk laces his fingers with mine, following me willingly, which only pisses me off more. Partly because I like it, and I really, *really* don't want to. And partly because this would be so much easier if he was just the bad guy instead of being Easton. *My Easton.* The man who's always been my knight in shining hockey skates and one of my favorite people in this whole stupid world.

I slam the bowl on the counter and turn toward him, fully prepared to tell him off, when he moves closer. "What are you—"

Easton's hands slide to my face, and goosebumps break out over my skin, making me forget what I was about to say as his clean crisp scent surrounds me. Calming me. "Telling my wife I missed her." He backs me against the fridge and covers my mouth with his. Soft and firm and impossibly perfect.

As much as I don't want to, I melt against him. Because it's Easton.

Hell, little cartoon fireworks might as well be exploding above my head with the amount of electricity firing off between us.

Then as quickly as it happened, the kiss is over, and he pulls back and presses his forehead against mine. "Hi," he whispers, and his minty breath fans my face.

"Hi," slips past my lips before I even have time to remember why I'm mad at him. Cautiously, I bring my shaking hand up to his face and stroke his cheek. "What are you doing, E?"

"Saying hello to my wife." There he goes again.

Yup. That helps me remember why I'm mad.

I drop my hand to his chest and push him back. "Don't get used to calling me that. I told you I want an annulment. The entire family is freaking out about it. My mom is ready to kill me, and I'd avoid Brandon at all costs if I were you. Scarlet's having a cow. ESPN is asking for an interview, and my brothers . . . Don't even go near my brothers."

"I'm not scared of your family, princess. I don't care what they think."

Why is his voice suddenly growly and sexy?

And why are my panties suddenly damp?

"You might not, but I do. Do you know how hard it's been to get my brothers and sisters to treat me like an adult? To stop looking at me like I'm a baby? I won a freaking gold medal, and I'm still not sure I was an adult in their eyes yet. But I was a whole lot closer before I got drunk and got married."

"Still don't remember that night, do you?" His thick arms cross over his chest.

I lift my face to his, and for a moment, I just stare at the man in front of me. The one who's meant everything to me for years. The one who looks like I just broke his heart. Suddenly, something cracks deep in my chest. "No. Not yet," I whisper.

The hurt is replaced by disappointment before Easton can mask it with cocky confidence. "You will."

"Maybe," I admit and realize there's a tiny nugget of hope in that truth.

Just as quickly as it vanished, Easton's grin is back in place. "You gonna let me stay here, wife?"

"It's Kenzie's home too." I try to hide the quiet tremor in my voice. "She said you can stay. And stop calling me that."

He takes one more step closer. "But I'm asking you, princess. If you say you don't want me here, I'll go."

I blink up at him and run my teeth over my bottom lip, hesitant to answer. Scared of the truth but completely unable to lie to him. "Of course you can stay here."

He reaches out with his thumb and presses it against my lip, freeing it from my teeth. "I knew you had it in you, princess."

"I said you can stay in my house. Not in my bed, hockey boy."

He cups my face in his hands and presses his lips to my forehead, sending a wave of warmth straight down my spine. "We'll see about that."

I dip out from under his arm. "Is that a challenge?"

"No, baby. It's a promise."

Why do I think those words are going to haunt me?

The Philly Press

KROYDON KRONICLES

BACK IN THE RED, WHITE AND BLUE

Looks like I spoke too soon, peeps. A little birdy told me that Easton Hayes is going to be wearing Revolution red, white, and blue the next time he laces his hockey skates. I guess Mrs. Hayes didn't like the idea of her playboy husband living it up in sin city while she looks for ways to spend her time, now that she's retired from skating. You know what this means? Baby watch is on! #babywatch #KroydonKronicles

LINDY

*B*y six a.m. the next morning, I've given up any hope of getting a good night's sleep and decide to give in to my desperate need for coffee. I throw my hoodie over my tank and pad downstairs, expecting Myrtle to wake up, ready to go outside. But my girl looks at me like I'm crazy as I pass her bed. Guess it's too early for her too.

Same, girl, same.

I tossed and turned all night, thinking about the fact that Easton Hayes was currently sleeping above me. *Ugh.* The images that single thought conjures makes me all sorts of squirmy. How is it fair that I finally spent a night sleeping in bed with Easton instead of just seeing his face on a phone, and not only do I not remember it, but apparently all I did was sleep?

Not that I'm mad we didn't actually have sex . . . I mean, this would be so much worse if we had. It's just—he's kissed me. *Twice.* And those kisses . . . My God.

Heat pools between my legs, just thinking about them.

Heat I ignore.

Nope.

No heat for me.

I refuse to accept any tiny little flicker of heat.

I cannot be turned-on by just the thought of Easton Hayes.

My husband.

I add the coffee beans and water to the sleek stainless-steel coffee maker and sit on a counter stool, staring at it. Willing it to work faster. I've got ice time at the rink in an hour, and there's no way I'm getting through the morning without a boost.

"Good morning, wife."

A chill skates down my spine as Easton joins me in the kitchen and drops a kiss on the top of my head like it's the most natural thing in the world. He's freshly showered, and his sandy-brown hair is wet, tousled, and smells delicious. Like white pine and citrus.

He moves past me, and oh my . . . Gray sweat pants hang off his lean hips and hug his thick thighs. And my mouth waters when I drag my eyes up his chest covered in a deliciously tight white t-shirt. This man is a god. No wonder I've had a crush on him for half my life.

He makes himself at home in my kitchen, like he's been here a thousand times, and grabs a mug out of the cabinet as he hums.

Hums.

Who hums at six a.m.?

Apparently, my husband does as he makes himself coffee. I watch, fascinated, as he fills a mug, grabs the Christmas-cookie creamer from my fridge, adds a heaping pour, then holds it out for me and waits.

I blink up at him, confused.

Easton leans in and licks his lips, and I swear my heart skips a beat. "I like when your eyes do that, princess. Your lashes get fluttery, and your cheeks are all pink and pretty. It makes me think about all the other things I can do to get you to flush that way." Easton's lips caress the shell of my ear as he whispers, "And I've gotta tell you, they're all a hell of a lot more fun than making you coffee."

He finally puts the coffee down on the counter next to me as I stare at him in shock. "Why do you know how I take my coffee?" I ask, stunned.

"Because I've made it my business to know everything about you."

"Oh." What the hell? *Oh?* That's the best I can come up with?

"Do you have plans tonight?" He doesn't wait for my answer before he starts rifling through my cabinets again.

"What are you looking for?" I ask.

"A travel mug."

I move to the other end of the counter and rise up on my tiptoes to grab him one of our insulated mugs. Then I freeze as his warmth envelopes me without his body ever touching mine. His arms move to either side of the counter, boxing me in.

I don't turn around, afraid of what he might see if I do.

Instead, I place the metal mug on the counter and take a deep breath.

"I promised Andrew I'd meet him at the rink tonight. He's still trying to get me to reconsider competing, and I keep telling him no. So now he has me watching his potential partners try out instead." I try to say it forcefully, but the words come out more like a whisper.

Is it possible to feel him even when he isn't touching me?

"Are you sure you're done competing? Won't you miss skating?"

It's the same question I've been asked too many times to count since the Olympics. But my answer has stayed the same. "I still skate. But now I skate for me, or when I'm working with my baby skaters. Now it's more fun, less stress. Andrew's just having a hard time accepting that."

"What time are you meeting Andrew?" Easton growls quietly against my ear.

"Seven," I breathe out and fight every instinct screaming at me to take one small step back. One tiny little move would close the distance between the two of us.

"Seven," he whispers and steps back, then grabs the travel mug. "Thanks, princess."

I turn slowly and watch Easton pour his coffee and screw on the lid. "Where are you going now?"

"I want to surprise Blaise and drive him to school before I meet with my new coach. Then I've got practice. First game's tomorrow night."

His words shake up a thought I hadn't considered before. "Have fun at practice, hockey boy. You might want to steer clear of Jace on the ice."

"Your brother loves me." He looks at me for a moment, losing a little of his bravado before he shakes it off and walks out of the kitchen calling out, "I'll see you tonight, princess."

Once I hear the front door click shut behind him, I sag against the counter.

It was a whole lot easier to be mad at him when he wasn't close enough to touch.

I blame the touching.

The touching leads to trouble.

The problem is . . . I think I'd like that kind of trouble.

Easton

Coach Fitzgerald stands from behind his desk at the Revolution arena and offers me his hand. "Glad to have you join the team, Hayes. It'll be nice to have you in our net for a change."

"Yes, sir. Thanks, Coach. Glad to be here."

He motions for me to sit. "I was surprised when your agent called and accepted our offer. You've been telling us to pound sand for a few years." He leans back in his chair and waits as the door to his office opens. "What changed?"

Well . . . Shit.

I should have seen that coming.

Max Kingston walks into the room. The man is certainly the master of his own universe. Even in his forties, he exudes power and strength unmatched by anyone else I've ever met. And right now, he's looking at me like I'm a problem he'd like to eliminate.

"Yeah, Hayes. What changed?" Max leans back against Fitz's desk with his eyes narrowed on me. Eyes the same color as his sister's.

I meet Max's stare head-on. "My reason for staying in Las Vegas changed. It was time to come home."

Fitz clears his throat, but Max ignores him. "And how long do you plan on staying?"

"His contract—"

"I'm not asking what his contract says," he interrupts Fitz, his glare never wavering from me.

"As long as my wife wants to stay here, this is where I'll be. With as close as she is to her family, I don't think she'll ever want to leave." I lean back in my seat and cross my leg. "I'd like to finish my career in Philadelphia, if I can, but that'll be up to you."

"Hurt my fucking sister and I'll make sure you never play another minute of professional hockey again. You won't be able to tend goal on a fucking development team in some no-name town in Canada when I'm through with you." His knuckles turn white from his grip on the edge of the desk behind him. "Do we understand each other?"

"I think—" Fitz tries to break the tension, but I refuse to back down because this moment is more important than hockey.

"Loud and clear, Max. But it goes both ways," I tell him as I stand from my chair. "Pretty sure I've already proven I'd die for your sister. How about you let her live her life like the intelligent, independent woman she is, and *you* try not hurting her? Because I'm pretty sure she'd be hurt if she knew you were assuming she couldn't stand up for herself."

Max takes a step forward, looking like he's ready to swing. "Watch it. I've been taking care of Lindy her whole goddamn life, asshole."

Coach slams his hand down on his desk. "Get out of here, Hayes. Practice is in an hour at the facility in Kroydon Hills. Don't be late."

"Coach—" I start, but he cuts me off.

"Go before Max kills you. I don't need my GM in jail and my goalie out of commission. Jesus Christ."

"Yes, Coach." I reach out and shake his hand over his desk, right next to Max, who doesn't move. "I'm looking forward to playing for you."

Fitz shakes his head. "Then get the hell out while you still can."

I nod once and walk out, without looking back.

Not exactly the welcome to the Revolution I was expecting.

EASTON

I may have just pissed your brother off.

LINDY

Which one? I have a few.

EASTON

The one who kinda owns me now.

LINDY

Max? Oh shit. He never gets mad. What did you do?

EASTON

Why do you assume I did something?

LINDY

Well . . . ?

EASTON

Okay. Fine. I married you. Apparently that was enough.

LINDY

Told you the family was furious.

EASTON

Becks isn't.

LINDY

He isn't?

EASTON

Nope. He trusts our judgment.

LINDY

Pretty sure he's the only one.

EASTON

Ever thought about standing up for yourself to your family?

LINDY

Ever thought about minding your own business?

EASTON

You are my business, wife.

LINDY

Don't you have practice or something, hockey boy?

EASTON

How do you know that?

LINDY

Lucky guess. Good luck.

The Revolution's practice facility is state-of-the-art. It was built a few years after the Kingstons bought the team, and from the looks of it, they spared no expense. The locker room is expansive. Wall-to-wall stalls are set up with our names above each one. It takes me a minute to find mine, and when I do, I run my finger over HAYES and close my eyes, knowing this was a long time coming.

Before I joined the Vipers, I bled red, white, and blue.

The Revolution was my hometown team, and I loved them.

When I was little, my mom used to wake up at the asscrack of dawn to drive me to practice before school. She'd never turn the radio on in the car. She liked to talk during our drives instead. We'd talk about everything, from whether I did my homework to what new show I was watching or what team I was playing that weekend. And we'd always talk about the day I'd play for the Philadelphia Revolution.

She was convinced it would happen.

Pretty sure it was her dream as much as it was mine.

I wish she were here to share it with me now.

She'd love this. The team and Lindy.

I drop my bag in my stall and change into my skates before the rest of the team gets here. I want to get a feel for the ice while it's still empty. I've already made a shit impression on my coach, but maybe I can do better with my teammates.

When I walk through the tunnel, I stop dead in my tracks. Looks like I'm not the only one who wanted a few minutes alone on the ice. Lindy is flying around the rink with Taylor Swift's "Wildest Dreams" playing.

She's fucking beautiful. She always is. But damn, when she skates, she takes my breath away. Her long blonde hair whips behind her with each new move, and as the song speeds up, so does she. She's gonna jump. *Shit.*

My breath catches in my throat as I watch her launch herself into the air. She gets three full rotations before she lands it, and her entire face lights up. Her arms go out, and she transitions into her next move and works through the rest of the routine, ending in a spin that gets tighter and faster until finally she picks up a skate and stabs it into the ice, stopping with gorgeous precision.

She's incredible. And she's mine.

She told me so, even if she can't remember it *yet*.

"You fucking stalking her now too, asshole? Gonna get her drunk, again?"

I turn when I hear Jace Kingston's pissed-off accusation but not fast enough to block the right hook he throws.

Fuck.

I stagger back a step, then right myself just as Lindy flies across the ice over to us.

"What the hell, Jace?" She shoves him back with both palms to his chest as she steps between him and me.

I move forward so I'm next to her, not cowering behind my woman like a fucking bitch.

"He was watching you skate like a fucking stalker, Madeline," Jace takes a step toward me, and she blocks him again.

"Jesus, Jace. A stalker? Really?" Her cheeks flame as she gets in his face. "Are you kidding me? I've dealt with a stalker, and that's not Easton." She shoves him again. "I'm a figure skater, Jace. People *watch* me skate. It kinda goes with the job, genius."

"*He* doesn't get to watch," he shouts with a finger pointed in my face. "He shouldn't even be here. He doesn't deserve to breathe the same fucking air as you after what he did," Jace yells back right in her face, and I see red.

I move in front of Lindy and wrap my arm around her, keeping her behind me. "I don't care who the fuck you are. You don't talk to her like that."

"Oh yeah, asshole?" He lowers his voice. "Why? Because she's your wife? What a joke."

"No, jerkoff. Because she's your sister, and she deserves more respect than that. If I wanted to defend her because she's my wife, you'd already be on your ass with a broken fucking nose. Because unlike you, I don't hit like a pussy. Then she'd be pissed at me too." Lindy squeezes the hand I have resting on her hip and drops her head to the middle of my back.

I don't move until she does.

When I drop my hand, she steps around me and stops directly in front of her brother. "Back off, Jace," she tells him more calmly this time. "And you might as well spread the word while you're at it. By the time you were my age, you had Cohen, and you and India were married. India, *who*, if I remember correctly, you were ready to marry after only knowing for a month. I've known Easton for half my life, you hypocrite."

"But—" he tries to jump in, but she shuts him right down with one look, and fuck me, but my dick gets hard as steel.

"Does your wife know you're acting like a caveman, Jace?"

"Madeline," he hisses.

"Don't *Madeline* me, big brother," she snaps back. "I know I'm the baby of the family, but I'm a grown fucking woman, capable of making my own choices."

"And he's your choice?"

Lindy looks up at me, her eyes blazing with fire. "He's my choice."

Holy shit. I'm pretty sure she's just saying that because she's pissed at Jace, but it's so damn close to what she told me Saturday night, and she doesn't even realize it.

Jace looks from his sister to me in disgust and shakes his head. "This discussion isn't over, Madeline."

"That's where you're wrong, Jace. There is no discussion. I don't need your permission." A few of the guys from the team start to file out of the tunnel, and Lindy pats Jace's chest.

That's when I notice her platinum wedding rings shining bright enough to blind me.

She's wearing them.

Ho-ly shit.

She's wearing them.

She turns to me and presses a kiss to my cheek. "Play nice and try not to piss him off more than he already is."

"Yeah, Easton. Stop breathing because it's pissing me off," Jace mumbles, and Lindy glares.

"Ignore him," she whispers and circles her arms around my waist. "And maybe try not to kill him during your first practice."

I wrap my arms around her and rest my chin on her head

while I watch Jace's lip curl in anger. "I can't make any promises, princess. But I'll try."

"Thank you." She lifts up and kisses my cheek. "I'll see you at home."

Jace growls, and I smile like I'm king of the fucking world.

She's wearing my rings, and she'll see me at home.

She might not remember, but it's a start.

LINDY

EVERLY

Hey! Anybody in the mood for sushi tonight?

KENZIE

Tonight's my late class. I'll probably just grab something on campus.

BRYNLEE

I'm in. We meeting there or doing takeout?

EVERLY

Have you seen the new bartender? We're meeting there.

GRACIE

I've got Nutcracker practice. Bring me home something?

BRYNLEE

Sure. Text me what you want.

EVERLY

Helloooo . . . Lindy? What about you?

LINDY

I'm about to meet Andrew at the rink. I'll grab something after.

KENZIE

Aren't you meeting Easton after?

LINDY

I didn't tell you that.

KENZIE

I have my ways.

EVERLY

Like on a date?

BRYNLEE

Scandalous. She's dating her husband.

LINDY

Whatever. It isn't a date. I just told him where I'd be. He might not even show. He had a rough day.

BRYNLEE

Because Uncle Max was an ass? Or because Jace decked him?

KENZIE

I'm sorry. WHAT?

LINDY

How did you hear that already?

BRYNLEE

I'm a physical therapist for the team. I hear everything.

LINDY

The entire team knows Max was an ass?

BRYNLEE

No. Mom filled me in on Max when she stopped by for lunch. But the whole team knows Jace hates Easton.

EVERLY

Your family can't keep a secret to save their lives.

LINDY

> Says the girl who posted my wedding photo on every social-media platform there is, then got so drunk that she didn't remember doing it?!?

EVERLY

I wasn't the only one who blacked-out that night, missy.

KENZIE

Can we go back to Jace hitting Easton?

LINDY

> It was awful. Jace is so pissed.

GRACIE

What did Easton do?

EVERLY

Did he get all possessive on your ass?

GRACIE

Did he throw you over his shoulder again like at the club? Because that was hot.

LINDY

> No. Better. He got all sorts of protective and defended me. He basically told Jace to stop treating me like a baby. It was actually pretty damn amazing.

GRACIE

And . . . ?

LINDY

> He may have called me his wife to my brother.

GRACIE

I take it back. That's hot.

LINDY

> It might have been a little hot.

EVERLY

You know a great way to thank him?

KENZIE

If you tell her to blow my brother, I might actually kill you while you sleep.

BRYNLEE

They're married now, Kenz. Pretty sure she's going to play with his peen.

LINDY

> OMG! Shut up. No talk of peen. I don't even remember marrying him.

EVERLY

Do you remember screwing him? Because it would be a crime against humanity to forget that.

LINDY

> I didn't have sex with Easton.

BRYNLEE

Did you have sex with someone else?

LINDY

> What? No!

GRACIE

Have you talked to a lawyer yet?

LINDY

> Not yet. I've got to find one.

BRYNLEE

You're related to one, genius. If you really want to get your marriage annulled, you'd have talked to Becket already.

LINDY

I'm kinda avoiding my entire family right now.

KENZIE

Or you don't really want the annulment.

EVERLY

I'm calling it now. You'll be kissing Easton when the ball drops on New Year's.

BRYNLEE

I'm with Everly.

LINDY

Mind your own business.

GRACIE

Ha. Like we ever do.

LINDY

Whatever. I've got to go inside now. I'll see you all later.

BRYNLEE

Say hi to your husband for us.

EVERLY

Get creative with the way you say it too. Like maybe on your knees.

LINDY

OMG. Goodbye!

THE KEEPER

The rink is nearly empty later that night while I sit and watch Andrew finish a routine with Cara, the young woman who's currently auditioning to be his next partner. She's been around the circuit for a few years. Most of us have. Competitive skating is a small circle. And to find someone at our level narrows it even more. Is she as good as me? No. But she's younger and has time to grow. She could easily surpass me with the right training.

As much as I know retiring is the right move for me, it's still the strangest thing watching my partner skate with someone else. Andrew and I've skated together for years. We can anticipate each other's next moves, and now we're essentially picking out my replacement.

Some days, adulting blows.

After their routine ends, I smile and wait for them to skate off the ice, pushing the twinge of unease aside, and give him the honest feedback he needs because that's what friends do. "That was great, guys."

"Thanks, Lindy." Cara beams and slides her skate guards over her blades. "That means a lot coming from you." She turns and smiles at Andrew. "I guess I'll see you tomorrow for the tryout with your coach then?"

Andrew nods. "Yeah, thanks. Same time tomorrow."

He watches her leave quietly before dropping down on the bench next to me. "Are you sure you're done? You don't want just a few more years before you give it all up?"

"We've talked about this already," I try to say sweetly, but I didn't come prepared for another guilt trip.

"Come on, Lindy. We're so good together." He throws his arm around my shoulder and leans his head against mine. "You've always been a competitor. Why stop now?"

"Andrew..." I groan and look up at him. "I will always be here for you, but it's going to be as your biggest fan, not your partner. You and Cara looked great out there. Her lines are beautiful."

"She's not you." His tone is sharp and sets me on edge.

I'm not in the mood for another angry person.

I've had enough of that this week to last me through the end of the year.

"No, she's not me. But I've been telling you for months that I'm done. If you're going to continue competing, you've got to find someone else to partner with. Cara is a great choice."

He tucks my hair over my shoulder. "What if I only want you?"

"Andrew . . ." I pull away.

"The lady said no."

Apparently, this day *can* get worse.

When I look up, Easton's intense gaze is locked on me. Is he . . . ?

He reaches out for me, and I place my palm in his without even thinking about it. A satisfied smile graces his lips as he pulls me to my feet. "You okay, princess?"

"Of course, she's okay," Andrew answers.

Easton growls before he rests a finger under my chin and lifts my face. "Ready for dinner?"

"I thought I was meeting you at home?" I ask.

"At home? Lindy . . ." Andrew looks between the two of us, then down at my hand. "He's a hockey player, Lindy. Come on. You can't seriously plan on staying married to him. They're Neanderthals."

"What the fuck—"

I press my palm against Easton's chest, then glare at both of them. "Stop. Both of you, just stop. I'm not an object you can fight over like two toddlers in a sandbox." They both look offended, but at the very least, they're smart enough to keep their mouths shut. "Andrew, I'm retired. That's not changing."

"Does this have something to do with *him*?" Andrew's disgust is evident as he looks between Easton and me, and I want to scream. I'm so over this stupid day.

Luckily for Easton, he doesn't say anything.

Nope. My husband simply wraps his hand around my hip, much like he did earlier, and squeezes, letting me know he's here. As if I could forget. As if the heat from his body isn't singeing my skin.

"I'm not even going to dignify that with an answer. If you'd like me to come to your tryout tomorrow night, let me know. I'm going home."

I move around Easton and ignore both men.

Andrew stands his ground where he is, but I feel Easton immediately move with me. He follows me through the doors into the parking lot before he grabs my hand. "Slow your roll, princess."

"Slow my what?" I spin around and shove him away. "Listen, hockey boy. Fighting is your thing, not mine. I don't like confrontation. I don't like arguments. And I really, *really* don't like violence. I've dealt with all three today. Now, I'm not saying they're all your fault, but they've all centered around you and our marriage, and if that doesn't scream something is seriously wrong, I don't know what will." I close my eyes, refusing to cry. Not now. Not in front of him.

Have I mentioned confrontation makes me cry? Because it does. And it's not pretty.

Easton's big, fat feet take two steps my way, but I throw my hands up. "Don't. Do not touch me."

"Lindy . . ." he whispers, and I feel horrible for the way those words just came out. "Tell me what you want me to do."

"Tell me why I married you? Tell me why you married me. Please," I plead. "Tell me why I've been fighting with almost everyone I know for a week, E. Tell me something. Make me understand what you're doing here in Kroydon Hills? Why take the trade to the Revolution? They've been trying to get you for years. Why take it now?"

Easton's steps are slow. Cautious. Like he's scared I'll bolt at any second. With one hand, he reaches up and cups my cheek. "Do you really not know, princess?"

I swallow down what little pride I feel like I have left and lift my eyes to his, blinking back the tears, and shake my head.

"It's always been you. The answer to all those questions is you, Madeline Kingston. It was you when we were too young for me to admit it was you. It was still you when I wasn't a good enough man for it to be you. It's been you every night in my dreams, when I'm not strong enough to save you. When you slip through my fingers and I lose you before I wake up in a cold sweat, unable to shake the image of you dying in front of me from my brain."

He rests his forehead against mine. "And for one night, it wasn't just you. It was us."

"Easton. I . . ." It might be the hardest thing I've ever done, but I force myself to take a step back. "I don't know what to say. It's hard to think when you're everywhere."

E shrugs. "Well I'll be gone for an away-game stretch over the next week. I leave the day after tomorrow. So I guess you'll have some time to figure it out."

Yeah. I guess I will.

"So when does he leave?" Everly asks as she pours herself a glass of wine. She and Brynlee just got back from the sushi place. Luckily for me, they grabbed me a California roll because I'm starving.

"I don't know," I tell them over a mouthful of food. "In two days, I guess."

Brynlee pulls her phone out and looks at something before looking back at me. "The team flies out at noon. We play in Atlanta that night, then have two more games before we fly home. Our next home game is in a week. I think it's our only stretch this bad all season."

"Oh." I push my food around on my plate with my chopsticks, not liking the sinking feeling this gives me. "Guys, what the hell am I going to do?"

"Girl, you've got to figure out what you want before we can try to figure out what you need to do," Everly tells me as she grabs Bryn's phone out of her hand and studies the screen. "Your man's games are on the other side of the country for a week. The man who just got himself traded. *For. You.* The one who just moved his whole life how many thousands of miles away. *For. You.* He knows what he wants. What do *you* want?"

I run my fingers through my hair and cringe. "I don't know what I want."

"Yes, you do," Bryn says sternly. "You're not some wishy-washy little girl, Lindy. You've wanted Easton for years. We're not blind. Everyone knows it, even if you're not sure. And guess what? You've got him. Now what are you gonna do with him?"

"How much sake did you guys drink at the sushi place?"

Everly laughs. "I told you the bartender was hot."

"Yeah well, it makes you mean," I tell them.

"Not mean. Direct," Bryn corrects me. "Listen, we love you. But it's not like we don't know Easton. It's not like we haven't watched the two of you together. You've danced around this for a long time. It's almost like what everyone says about your mom and Brandon. Don't wait as long as they did to figure it out. If you've got feelings for him, figure out what you want now. And if you don't . . ."

My head snaps up. "If I don't, what?"

Everly cocks her head. "If you don't, then set the man free, and don't come crying when someone else snatches him up. Because drunken wedding aside, that man is a keeper. He's gorgeous and tall with big hands and big feet, so my money's on a big—"

"Don't," I snap. "Don't go there."

"Fine," She sips her wine and attempts to hide her smile. "But I'm just saying, if you let him go, you better be sure that's what you really want because you won't get a second chance."

I know she's right. I've hated knowing he was hooking up with women over the years.

I always dreaded the thought of him bringing someone home to meet the family.

"I hate you both," I tell them, and Brynlee laughs while Everly just shakes her head.

"No, you don't. You love us," Bryn says.

"Whatever. I'm not ready to let him go," I admit and stuff another piece of California roll in my mouth, completely unwilling to say anything else.

"Oh, Lindy. That's not good enough." Everly snatches a piece off my plate.

"Hey, you ate already." I poke her with a chopstick.

"Hot bartender, remember?" she argues. "We don't eat in front of the hot bartenders. I'm starving." She stands and looks down at me. "Do you want him? Or do you just not want anyone else to have him, Linds? Because if it's the second, you're a better person than that."

I grab her glass of wine and finish off what's left of it. "Fine. I want him. I've always wanted him. But not like this."

"Then how do you want it, princess?"

Everly, Brynlee, and I all shriek, and Easton looks at the three of us like we're absolutely insane.

Everly snorts. "Jesus Christ, man. Make some fucking noise when you walk in the house."

"We're just gonna . . ." Bryn grabs Everly's hand and tugs her out of the kitchen but not before Everly manages to grab the black to-go container from the table.

I watch them both run away to hide and silently wish they could take me with them. "How much did you hear?"

He looks so damn good in worn blue jeans and a white thermal shirt stretched tight across his chest. Strong and sexy. This man is every fantasy I've ever had, and he's standing right in front of me. But I don't handle being trapped all that well, and I'm feeling cornered right now.

"I heard you say you want me but not like this. Tell me what you want, Lindy. We've shared so much over the years. You can't start holding back on me now." He leans back against the counter and crosses those arms I love over his chest, and his shirt stretches so damn tight around his biceps. My God. It should be illegal to look that damn good.

"Easton . . . Haven't you ever noticed I never talked to you about this? Not about you. Not about any guys. I've never crossed that line," I tell him because I don't know how to give him what he wants.

"I was kinda hoping it was because there weren't any guys."

My blood burns with indignation. "You've got to be kidding. You used to FaceTime me with women in your bed, but you didn't want me dating? Seems awfully hypocritical, doesn't it, E?"

"I never said I was a saint, princess. But think about it for a minute. No matter who I was with, I always ended up calling you. Doesn't that tell you everything you need to know?" His hazel eyes are so damn intense, I'm not sure how much I can take.

"You don't actually think I liked it when you'd call me with a woman in bed next to you, do you?" I demand.

"Tell me why it bothered you," he pushes back.

"Because..." I stammer.

"Use your words, Madeline."

"Ugh... You're infuriating." I push back from the counter stool and march across the kitchen to him. "You want to know why it bothered me?"

"Yeah, I do, princess." He stands tall as I stop in front of him, ready to scream.

"Because I was jealous." I throw my hands up. "There. Happy?" I yell. "I was jealous they got to be with you like that. In a way I was sure I never would. I was so goddamn jealous because I wanted to be them."

Easton leans in slowly, careful not to touch me. "None of them could ever be you."

"But they had you in a way I couldn't, and I hated that," I admit, a little quieter.

"Then tell me why you were so pissed when we woke up married," he pushes again.

"You're really going to make me say this?" I ask as my voice shakes from frustration mixed with embarrassment.

"Yup."

"Because I've wanted you for years, Easton Hayes. And I finally got you, but it was because we were drunk. You never gave a shit about me sober. But I got drunk and flirted with someone in front of you, and you thought, *Fuck that, I'll marry her.*" My stomach twists as the words keep coming. "I woke up and was married to the one man I've dreamed of marrying, and I couldn't remember any of it. You don't even love me. You—"

"Stop." He slides his hands up my neck and cups my

cheeks. "Forget the fact that we're married for a minute. What do you want from me, Lindy? Do you want me? Want us?"

I lift my eyes to his, and the softness there just about kills me. "Have you listened to anything I just said? Of course I want you. I've only ever wanted you, you idiot. But I'm scared, Easton. You've been my protector for so many years. What if this doesn't work out? I can't lose you."

"You'll never lose me." His thumb caresses my skin. "I'm in this for the long haul, princess."

"But you can't know that," I fight back, desperate to get him to feel my fear and take it seriously. "You can't be sure."

"I've never been more sure of anything in my life, princess. It was always supposed to be you and me. The timing was just never right before." His words are whispered against my skin, wrapping around me like a safety blanket.

"Easton . . ." I plead.

"Give us a chance, Lindy." He presses his lips to my forehead.

Ugh. How am I supposed to think when he's doing that?

A chance.

Can I do that?

Can I give him that?

Can I let go of that fear?

"A chance?" I whisper back, and Easton tilts my face up to his. "Fine. If you want a *chance* . . . If you want to even think about staying married to me . . . We need a reset. We need to date. We need to go back to the beginning and redo all the steps we skipped. You're going to have to woo me."

"Woo you?" A smile spreads across his perfect lips. Lips I desperately want to trace with my tongue.

"Yes, Easton. If you want to be my husband, then you've got to start with the baby steps and win my heart." I don't bother telling him he's had it for a lifetime.

"Woo you," he repeats again.

I lick my lips and nod.

"I can do that," he whispers, and just when I think he might kiss me, he runs his thumb over my lip. "Game on, princess. Prepare to be wooed." Only instead of kissing me, he smiles a devious smile and walks away.

"Where are you going?" I ask, shocked . . . again.

"Oh, baby. I've got work to do."

My head spins from whiplash as I watch him head for the steps and wonder where, exactly, I lost control of the night. And how, exactly, I ended up here.

The Philly Press

KROYDON KRONICLES

TROUBLE IN PARADISE?

Rumors are running rampant for the Revolution, and Easton Hayes is at the center of them. Is there already trouble in paradise for Kroydon Hills' favorite *It* couple? I feel like we need a nickname for them, but at this rate, they may be over before they even started. Sources tell me the new Mr. Madeline Kingston is fighting with her brothers. His hockey captain in particular. Looks like Jace Kingston just gave Easton a black eye. Could it be that he's already mistreating Madeline? Stick around and see.
#KroydonKronicles

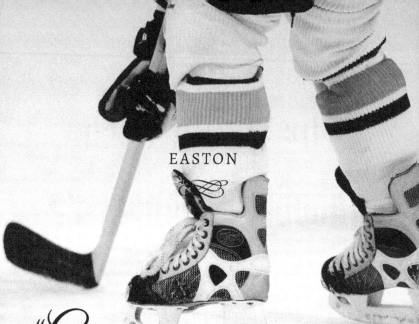

EASTON

"Glad to have you back home, brother." Pace squirts enough ketchup on his plate to drown every french fry in the place as he sits across from me at The Busy Bee Café on Main Street.

"I've been home for a few days, man. It's not my fault you were flying around the country visiting your other clients."

He stuffs another fry in his mouth. "It's not like you're my only client, Hayes. You know you can be a real diva when you want? Maybe I should start calling you princess instead of your wife. How's it going on that front, anyway?"

"We're dating."

"The fuck?" he asks, confused.

"She's agreed to hold off on the annulment for now. So I'm counting that as a win, and we're going to date."

"But you're already married . . . Wait. Don't they say blow jobs stop once you're married? Maybe this is a good thing. Get all the head now, while you can." The asshat laughs like a fucking clown, and I kick him under the damn table.

"Pace, man, were you dropped on your head as a baby?"

"Pretty sure I was once or twice. I blame my brother, Hunter. It's his fault." He takes a bite of his BLT, then opens his mouth full of food. "Anyway, how's the new team been?"

"Dude, fucking chew before you talk. Most of the guys are okay. Jace is gunning for me. So that's awkward as fuck. Max threatened me on my first day, but I haven't seen him since. Oh, and I like Coach Fitz. He seems like a good coach, so far." Fucking Kingstons.

"You sure she's worth it?"

I don't bother answering, and Pace smiles.

"Yeah. That's what I thought. Have you seen Juliette?"

I finish my burger and wipe the grease from my mouth. "Nah. She and Becks are down in DC for a few days. Some White House Christmas party. I've talked to Becks but not Jules."

"Chicken shit," he busts my balls. "Did you get the new place situated?"

I shake my head. "I ordered a bunch of furniture that's gonna be delivered while I'm gone. Kenzie's gonna get it all set up for me. I guess I'm gonna move in when I get back."

"You guess?"

"Yeah, man. I can barely get face time with Lindy now. I don't know how much worse it's gonna be when I move out."

"I hope your girl's worth it."

"My girl's always been worth it, dick," I tell him and throw my napkin at his face.

"I knew she was your girl."

Yeah, deep down, she's always been my girl.

Lindy manages to avoid me all day.

She was gone by the time I came down this morning, and she's not home when I get home from practice that night.

My whole fucking body aches from the extra conditioning Fitz threw at us when Jace and I got into an argument in front of the team during practice . . . again.

Neither of us touched a puck for the rest of the session.

It was all suicides and sprints until we were both puking in trash cans.

My mood is shit, and I leave tomorrow.

The stairs outside of the loft creak, and I look up just before Kenzie pops her head in.

She looks at the bruising on my face and gasps. "Oh my God, E. Your face." She rushes over to me like I'm a kid who just got hurt on the playground.

"It's fine, Kenz." I shrug her off. "It happens all the time." As soon as the sentence is out, I know I shouldn't have said it.

Her eyes narrow, and her mouth tightens into a line. "Okay, am I supposed to be okay with you getting hit all the time? Aren't goalies supposed to be out of most of the fights, Easton?"

"It's professional hockey, Kenz. Shit happens."

"Oh yeah? You're supposed to get into a fistfight with your captain?"

"It wasn't a fight. Jace is pissed because I married his little sister." Looking at my own little sister, I can understand why he's mad. But it doesn't change anything.

She plants her hands on her hips. "Did you hit him back?"

"Sometimes words land harder than fists, Kenz."

She blinks up at me, shocked. "That may be the smartest thing I've ever heard you say, Easton Hayes. But I'm still going to kill Jace Kingston."

I laugh and tug her hair. "You're not gonna kill Jace, sis."

"Fine," she huffs. "I'll do one worse. I'll tell Juliette, and she'll do it for me."

She pulls her phone from her pocket, and I yank it from her hand before she can sic Jules on Jace. I don't need anything to make things worse for me with my new team. They're already not sure what to make of me, judging by today. They loved Jonesy, their former goalie. But whether they want to accept it or not, his injury from two weeks ago was career-ending. And the backup ain't cutting it. Jace is the fucking captain, so they're gonna take their cues from him, which means, right now, it's not looking too good for me. "No, you're not," I tell her and stick the phone in her pocket. "I don't need Jules fighting my battles for me."

I bend my knees to bring myself eye to eye with my sister. "I'm not even sure Jules would fight them right now anyway. She's pretty pissed."

"No, she's not." When I lift my head in disbelief, she laughs. "*She's not*. Well, not exactly *mad* you two got married. She's hurt you didn't tell her. She's coming home from DC tomorrow morning."

Kenzie smiles when Myrtle makes her way into my room and rubs herself against Kenz like a cat instead of a fifty-pound bulldog. "I swear to God, if you tell Jules I told you that, I'll sic Jace on your other eye."

I touch the pale bruising from when he hit me the other day and cringe. Not a great week. "I won't say anything."

She huffs and looks at the clothes laying on my half-packed bag that Myrtle just made herself comfortable on, and her body goes rigid. "Where are you going?"

"I've got an away-game stretch. I'll be back next week. I told you yesterday. It's why you're getting the condo ready, remember?"

"Yeah. I just hate that you're leaving. You just got here."

Kenzie and I aren't the greatest with people leaving. That's what happens when your perfectly healthy, young mom dies of the fucking flu.

"It's the team's schedule, Kenz. I can't control it. But I'm coming back. I promise." I wrap my arm around her and squeeze. "I signed a five-year contract. You're not getting rid of me that easy."

"Promise?" she whispers, and I suddenly feel like a dick for staying away so long. I was so wrapped up in my own shit, I didn't think about how being away would affect her.

"Promise." We sit quietly for a minute, then I decide to test the waters. "So . . . any chance you'd want to help me?"

"With what?" She perks up.

"Lindy."

Her smile stretches across her face. "Let's see . . . do I want to help my brother win over my best friend so she can literally be my sister? Hmm . . . What do you think?"

"I think I need to romance her."

"Yes, you do," Kenzie laughs before she drops her head on my shoulder. "I guess there's hope for you yet."

I sure as hell hope she's right.

I haven't been a light sleeper in, well, ever. Even as a kid, I heard every noise. Every beep of a horn. Every conversation my mom thought she was having in private because her kids were sleeping. Everything wakes me up, and that's if I'm even able to fall asleep in the first place.

If I'm lucky, I get four hours a night. It's not healthy, especially for an athlete. But I'm used to it. I've adapted. So when the stairs leading up to the loft creak at two a.m., I'm wide-awake and looking at the open doorway, waiting to see who's coming. I'm half expecting my new best friend, Myrtle, to be looking for a warm spot to crash when I see Lindy hesitate at the opening.

"Easton," she whispers, and my stomach drops because I know that tone.

"You okay, princess?" I force myself to stay in bed. The last thing I want to do is push her right now. She came up here. The ball is in her court.

"No," she tells me softly but stays frozen at the door. Her long hair is a tangled mess around her shoulders as she stands in front of me in the white t-shirt I helped her put on Sunday morning before she stormed off. "I had a dream and picked up my phone to call you . . . but you're already here."

"I'll always be here. Come here, baby," I whisper.

Lindy pauses, then slowly tiptoes over to the bed where I lift the blanket and make room for her next to me. She looks down at the mattress with such hesitancy that until she gently climbs in and fits herself against me, I'm not completely convinced if she'll get in or go back to her room. "I haven't had one this bad in a long time," she admits so quietly, I barely hear her before she lays her head against my chest. "I could feel the barrel of the gun pressed against my head. It was so cold. And he just kept saying over and over again that he was going to make Mom watch as he shot me. *Make her watch as he killed me.*"

The tremble in her voice breaks me because that's not a nightmare.

That's what we lived through.

"You're safe, princess." I wrap my arm around her and press my lips against her head. "We got out of there. He'll never hurt you again." Sometimes I wish I'd been the one who killed him for what he did to her.

She grips my shirt in her hand and shakes. "In my dreams, you always save me."

"I always will." I run my fingers through her soft hair and shift my hips away. Pretty fucking sure she doesn't need to know my cock is ready to rip through my boxers, it's so damn hard just from the feel of her bare legs pressed up against mine. "Sleep, Lindy. You're safe. He can't hurt you anymore. No one will ever hurt you again. I'll never let them."

We've never done this.

Lie together like this. Bodies tangled together. Awake. *Sober*.

I carried her to my hotel room the other night and laid her down on her side after she passed out. Then I watched her sleep for hours, memorizing this woman.

The sugary sweet scent of her hair tickles my nose as I inhale her with each breath. Her delicate curves mold to me like she was made to fit against my body. Because she was always meant to be mine.

The quiet of the night is almost deafening as I listen to her breathing even out. "Did you mean what you said before, E?" The weight of her words hangs heavy in the air.

"I said a lot of things recently. I meant them all. But you're gonna have to be more specific if you want a specific answer." I get the feeling the importance of my answer matters too damn much to chance the question being wrong.

"Yesterday, you told me it was always *me*. Did you mean it?"

"More than I've ever meant anything in my life, baby," I whisper back without a single fucking second's hesitation because it's maybe the most honest thing I've ever told another living person.

Lindy is quiet for a long few minutes that stretch on like fucking hours. So long I wonder if she's fallen asleep. And I think she may have until she moves her head just enough to look up at me. Her long lashes kiss her cheeks as she blinks away tears. "It's always been you too, Easton."

She drops her head back down to my chest and wraps her arm around my waist. "It's only ever been you."

I don't move a fucking muscle after that.

Within minutes, Lindy's asleep in my arms.

When I open my eyes hours later, I'm more rested than I've felt in fucking years. I'm also sweating my balls off because my wife is wrapped around me like a vine and a snoring bulldog is pressed against the inside of my knees. She legit sounds like a cartoon dog snoring. Like Scooby-fucking-Doo. What the hell?

Lindy's breathing catches as her body goes rigid, and I almost laugh.

Guess we both slept well last night.

"Good morning, princess."

She slowly extricates herself from me, pulling her leg back from where it's thrown over mine, rubbing up against my cock in the process. "Umm . . . At least I remember last night this time." She buries her face in my chest. "I can't believe I slept here."

"Don't be like that. I fucking loved having you next to me. That's the best night's sleep I've ever had. Even Myrtle didn't wake me up." I sit up and lean back against the pillows, bringing her with me, loving this sleepy, soft side of her. "You're my wife. You're supposed to be in my bed."

"Technically, you're in *my* bed," she teases, and a pretty little flush creeps up her face. "I guess you can stay as long as you need to."

"You don't need to worry about that. I lined up a place to rent yesterday."

I watch as her blue eyes dim. "You're moving out? Already?"

"Didn't think you really wanted me here," I tell her as I run my fingers up and down her spine.

"I . . ." she flusters. "I didn't mean to make you feel that way."

I lift her chin up and bring her eyes to mine. "I'm twenty-seven, Madeline. I need my own place. I can't crash on your spare bed forever, and even if I could, I don't think I'd survive living with four women who aren't my wife."

"Where are you moving?" she asks, seemingly disappointed and a little pouty. And damn, doesn't my dick like that look.

"The two-bedroom downstairs. The one next to Maddox and Callen."

Her expressive eyes grow wide as she blinks before she bites down on her bottom lip. "Oh."

I press my thumb against that pouty lip, freeing it, and my girl sucks in a sharp breath. "Oh? That the best you've got?"

Her lips wrap around my thumb and kiss it gently, and I groan as the feeling shoots straight to my dick.

"I need my own place, *wife*, because like you said last night, I need to date you. And that means I need to pick you up at your door and bring you home afterward."

"You do need to date me, hockey boy." A devilish smile spreads across her face. "What do you plan on doing from your own place that you couldn't have done from mine?"

"Baby, I'm gonna make you scream my name so loud, your roommates wouldn't be able to look at you again without turning red."

She runs her hands under my t-shirt, dragging the tips of her fingers along my abs, and my muscles contract. "Madeline," I warn as she plays with the waistband of my boxers. "I'm trying to be a gentleman here."

"Pretty sure *gentleman* went out the window with me screaming your name, E."

The way she looks at me . . . *Fuck.*

I wrap my hand around her long neck and run my thumb over her thrumming pulse. "Lindy . . ." My other arm wraps around her back, and I brush my mouth over hers, groaning when her sweet taste explodes on my lips.

Lindy moans, and electricity zings between us. My cock pushes against my boxers, and I pull her into my lap, dragging her closer. Needing more. So fucking much more.

But knowing it's too soon for that.

This girl deserves romance and roses and candles.

She deserves the fairytale.

She *deserves* everything.

Trailing my mouth along her jaw, I taste her skin and touch her the way I've wanted to for fucking years.

The sexy sigh she exhales as she slides her hands up my chest and drags her nails against my bare pecs is intoxicating. My dick presses against her soft thighs, wanting in on the action.

As if knowing what I need, Lindy slides herself along my cock and moans as she leans into the kiss. Her tongue testing the waters. Tentatively touching. Learning. Exploring. Igniting every fucking nerve-ending like a wildfire that's just starting to burn.

One you know is going to burn out of control.

Her nails dance across my nipples, and I fucking growl as I run my tongue down her neck, stopping to suck her racing pulse.

Lindy clings to me, grinding against me like we're two teenagers dry humping, afraid to get caught. The only thing separating us are the scrap of silk of her panties and my boxers.

The heat of her pussy tempts me to take this further than I know we should.

Not yet.

"Lindy . . ." Unfocused eyes stare back at me, scorching my soul.

"I'm sorry" she whispers and crushes my heart with her words.

"Baby, you've got nothing to be sorry for. You were right last night. We skipped a couple of steps. I want to take you out and treat you like the princess you are. I want to romance you before I worship every inch of your body for the first time. You deserve that. You deserve everything."

She drops her forehead to my chest and shakes. "I have to tell you something, Easton."

I run my hands over her hair and cup the back of her head. "You can tell me anything."

Her shoulders lift and fall with a strong breath—in and out—and I brace myself for whatever the hell she's about to throw my way because it can't be anything good.

Her Bermuda blue eyes lock on mine, and she runs her teeth over her lip. "I've never done this before."

"Done what?" I ask, confused.

"Any of this?" she whispers but doesn't look away. "I mean, a little under the shirt action years ago. But that was all . . . before. I've never really trusted anyone enough since that night. Not enough to make myself *that* vulnerable," she says softly as she shakes.

I cradle her head in my hands. "Breathe, Lindy," I whisper against her lips. "Do you trust me?"

"With my life," she tells me, and my chest swells.

"That's all that matters. There's no rush for everything else. I've got to woo you first, right?"

Lindy smiles and presses her lips to mine. "You're going to make me regret using that word, aren't you?"

"Oh yeah, baby. But that's gonna be the only regret. I promise."

"Okay." She rests her head on my chest, and I know I just made the biggest promise of my life. Now I've just got to make sure I can keep it.

LINDY

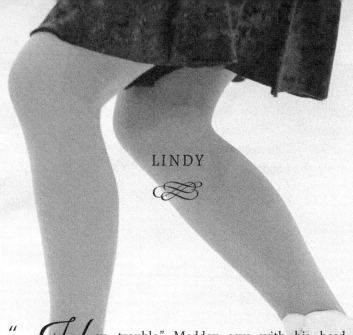

"Hey, trouble," Maddox says with his head buried in our fridge while he looks for something.

This is what he and Callen do. They let themselves inside our place, eat our food, drink our coffee, and occasionally steal our toilet paper. Spoiled babies. I'm not sure why we gave them a key.

"What are you doing here?" I grumble and steal his coffee from the counter, then spit it back in the mug. "How do you drink this black?"

He turns with a plastic to-go container in one hand and tugs the coffee away from me with the other. "Whatever. Why are you home? Don't you have lessons or some shit?"

"Oh, I'm so sorry I'm here. *In my own home.*" I pull myself up to sit on the counter and pour my own cup of coffee, then shove him back toward the fridge with the toe of my sneaker. "Where is everybody?"

"What's with the kick?"

"Grab the Christmas-cookie creamer for me, please," I

push, and at least the little mooch gets it for me without being a pain in the ass.

He adds a ton to my coffee, grumbling about how I'm gonna be a diabetic one day. "Where's everyone else?" I ask again before I look at the clock and realize just how late it actually is. *Damn.* I've got to get moving or I'll be late for the lessons I teach at the rink.

"Brynlee just left. The rest of the girls were gone before I got here." Something must catch Maddox's eye because he stops and leans back to look up the stairs. "Got something to tell the class, trouble?"

"Nope," I answer and sip my coffee, acting completely oblivious to the sound of Easton coming down the stairs.

Maddox looks between us when E walks into the kitchen and drops his bag on a stool. He reaches around me for the coffee with a handsome smirk on his face, and I feel my cheeks pink.

"Dude, stop. You two might as well have cartoon birds flying around your heads. What the hell?" Maddox groans, and Easton and I both laugh.

"You into cartoon birds, Mad?" I taunt. "I mean, your dad calls your mom Snow White. Is that like a kink for you now?"

Maddox's face turns bright red before he points at me. "You're fucking gross, trouble. My mom's a saint. She doesn't have a kink," he argues, and I can't stop the ridiculous laughter bubbling up.

"Oh my God. First, your mother, who happens to be my sister, is no saint," I practically double over, unable to breathe because I'm laughing so hard. "But even better, I was talking about *you*, you stooge. Not Amelia. I said *you* had a kink. But I mean, if you want to think about what your parents do in bed, you do you, boo."

I hop off the counter and pat his back. "See you later, madman."

"I hate you, trouble," he calls after me as I grab Easton and walk him to the door.

"He loves me," I tell E before I run a hand down the front of his hoodie. "It's December in Kroydon Hills, hockey boy. You may need to get a coat. Maybe even a hat," I add dramatically.

"Want to go shopping with me when I get back? I'll need all sorts of stuff for the condo too." He cups my face with his hand, and goosebumps break out over my skin. "Maybe you'll even let me take you to dinner."

"I'd like that."

He kisses my forehead as his thumb caresses my jaw. "See you soon, princess."

"Try not to kill my brother while you're gone, please," I whisper, half serious.

"As you wish," he tells me, and then he's gone.

Wow. I wasn't expecting the wave of sadness that washes over me.

A week ago, I was furious with him, and now, I kinda don't want him to go.

"As you wish," Maddox snickers as he tries to sneak by me, and I smack the back of his head before he gets through the door. "Smart people are scared of me," he taunts.

"*They* don't know you peed the bed until you were six." I slam the door shut behind him and giggle.

One day, he's going to run the Philadelphia Mafia.

But right now, he's just a pain in my ass.

Easton

THE KEEPER

I'm sitting on the team plane with my earbuds in and my eyes closed when I feel someone sit down in the seat beside me, so I crack an eye open, half expecting it to be Jace. It's not him, but it's not much better. His co-captain, Boone Dornan, is staring at me, waiting. For what, I'm not sure.

I lift my chin. "Boone."

"Hayes. You feeling good about going against Atlanta?" he asks as he adjusts the seat.

"I've shut them down before. I'll do it again."

"We do things as a team in Philly. You're not out there alone, you know." I'm not sure if this guy is serious or if he's giving me shit for being cocky.

"I hear you. You don't have to worry about me. I've got no problem being a team player."

"About that . . ." he drags out. "What the hell is going on with you and Kingston's little sister? Because I've known him for a really long time, and he's never been this big of a dick before."

What the fuck?

"She's my wife."

Jace stands up from two rows in front of me and turns around, looking like he's ready for round two. "She's not your fucking wife, asshole. Drunken Vegas weddings don't mean shit," he yells as the guy next to him holds him back.

It's gonna be a really fucking long week.

*O*nce we get to the hotel in Atlanta, our room assignments are handed out, and keys are distributed before we're given the schedule for the rest of the day. Some of the guys go to the conference room to grab food before we're needed back on the bus, but I head right to my room. I need to get my head on straight before this game, and I can't do that surrounded by a bunch of noise.

I played for the Vipers for a long damn time. My teammates and I had a rhythm on the ice. I could anticipate the plays.

Here, I haven't had one practice with the Revolution where Jace Kingston didn't spend most of it trying to get a slapshot off at my head.

News of the trade has been everywhere.

Interest is at an all-time high, and I don't want to fuck this game up.

When I walk into the room, I drop down on one of the two queen beds. At least I'm sharing a room with Boone and not Jace. Apparently, Fitz thought it would be good for a captain to show me the ropes. More like it would look good to the rest of the team if both captains didn't hate me.

My phone buzzes next to me, and I make the mistake of looking at it in case it's Lindy. It's not. Instead, Jules's name is flashing.

Might as well get this over with.

I swipe my thumb across the screen, accepting her FaceTime. "Hey, Jules."

"Look at that. You *do* know how to answer your phone."

I groan and sit up. "Come on. I stopped by the house twice before you went to DC. It's not like I was avoiding you. How was I supposed to know you wouldn't be home either time?"

"Well, if you'd have bothered to answer your phone, I could have told you I wouldn't." Her face softens. "I was hoping we'd get to talk before you flew out for the Atlanta game."

"Can we not do this now? I've got to get ready for the game."

Her face falls, and she plays with the pendant hanging from her necklace like she always does when she's upset. "Can I ask you something?"

"Go ahead," I grumble.

"Why do you think I'm mad at you? Haven't I supported everything you've ever wanted to do? Even when you were little and I was living in Europe? Whenever your mom would call and tell me you were trying out something new, I was always the first person after her to cheer you on. I'd call, and we'd talk for hours about how excited you were. We don't talk like that anymore, E."

"I'm not a little kid anymore, Jules." No matter how much I sometimes wish I could go back to when I was and my mom was still alive.

"No. You're not. You've grown into a man I'm so incredibly proud of. A man any woman would be lucky to be loved by. So I've got to ask . . . do you love her, Easton? Did you marry Lindy because you love her or was it a drunken Vegas mistake?"

"That's between Lindy and me, Jules." I try to keep my tone calm, but I'm not sure I manage.

"Easton—"

"Do you think I'd marry Madeline Kingston if I didn't love her?" I bite back, aggravated she'd even ask me that. "What the hell have I ever done to make you think so fucking little of me?"

"No," she whispers. "I don't. But I wouldn't know that because you haven't talked to me about it. I didn't even know you got married. I had to find out when Ashlyn called me, losing her mind. The Kroydon Kronicles seems to know more than we do. So tell me, should we all be on baby watch like they're reporting, Easton?"

"No, Juliette. Lindy's not pregnant." Not that I hate that idea. "And, okay, I'll give you that one. You shouldn't have had to find out that way. But we were still figuring things out," I tell her.

"What's there to figure out?" She brings the phone closer to her face as she moves around her office.

"Well, for starters, she doesn't remember the wedding. She and Everly thought pounding champagne would be a great way to celebrate afterward, and Lindy's not much of a drinker. So the night's a blur for her."

"Oh . . . That's not good."

"No, it's not. But she wasn't drunk when she married me. Neither of us were. We'd had a few drinks, but we weren't that bad. Then there's the fact we kinda went about all of this *backward*. We're trying to reset."

Jules sits down at her desk and switches me over to her laptop. "What do you mean reset?"

"We didn't date. We went from what we had"—which I'm not about to discuss with Juliette—"to married. She wants me to woo her."

"Oh, for the love of God. Seriously? She said *woo*?"

"Yeah. Why?" I ask, not liking the sound of that.

"Let's just say she's more like her brother than she probably even realizes. I hope you've got a plan because these Kingstons can be a real pain in the ass."

"I do," I tell her just before the door to the hotel room opens, and Boone and Jace walk in. "I gotta go, Jules."

"Hey, Jules," Jace calls out, sugary sweet when he hears her name.

Too bad I want to punch him in the face.

"Jace Kingston," Jules yells. "You little shit. If you so much as—"

I end the call before she can finish her sentence and pocket my phone.

Guess Jules wants to punch him too.

Boone's smile grows as he looks at the phone. "Dude. Is your mom the hot supermodel?"

"No, asshole. She's my cousin," I growl.

He smacks my back and drops down onto the bed where I was just lying. "Cool. Guess we're at the nickname stage of our relationship already. Can I call you little Kingston?"

"I'm not a Kingston."

"He's not a Kingston," Jace says at the same time as me.

"Chill, guys. I was kidding. It would be pretty gross if you were, and you still married his little sister. I mean, I'm pretty open-minded, but that crosses a line. Okay, how about I call you cradle robber?"

"Boone," Jace warns.

"I kid. *I kid.* Fine. I'm just trying to lighten the mood. You two have to put this shit behind you before either of you get on the ice tonight. We can't have our captain and our goalie ready to fight each other instead of the other team."

My damn phone vibrates again, only this time it's Lindy's face on my screen. It's a selfie she took of us kissing at the chapel the other night, then set as my contact for her. And instead of her name, *Wife* flashes there. Damn. I wonder what she'd do if she knew she did that.

Jace sees it too and grumbles, "Go on and answer. I'll wait."

"Yeah, I bet you will." I take a few steps away and answer, "Hey, princess."

"Fucking hell," Jace bitches.

"Hey, hockey boy." Her smile lights up her face. "I got my present. Thank you."

"Real original. Already resorting to buying her love?" her asshole brother trash-talks.

"Was that Jace?" Lindy asks, her smile gone.

"Yeah. Boone's here too," I tell her and watch her eye twitch.

"Can you put Jace on the phone please?"

He takes the phone from my hand, his chest puffed up like a fucking peacock who just got exactly what he wanted. "Nobody should have to buy your love, Lindy."

"My husband sent me his jersey so I could wear it tonight while I watch the game, you big bully. He knew I wouldn't want to keep wearing your name and number, not now that I know what a jerk you are."

God, I love this woman.

"Kingston is your name too, kid," Jace chastises.

"Not for long. Maybe I'm going to take Hayes as my last name. It sure looks nice on a Revolution jersey. Nicer than Kingston. Now put my husband back on the phone and go away unless you want to hear me tell him exactly what I'm going to do to him when he gets home, wearing this jersey and absolutely nothing else."

Jace's face grows red with mortification mixed with outrage, and Boone's lights up with excitement. "Dude, yes. I wanna hear."

"Shut the fuck up before I break your face too, Boone." Jace shoves Boone to the door before I even get the chance to lose my shit.

"Bus leaves in ten minutes, Hayes. Don't be late," Jace warns before he slams the door shut behind himself.

"Princess, you can't say stuff like that in front of your brother."

"Yeah, but his reaction was worth it." She moves in front of the mirror in her hall and angles the phone so I can see my name and number eighty-eight on the back of her jersey. "But I really do like my present. Not that I'm some needy little bitch who wants you to buy her things. This was just very thoughtful."

"Don't thank me too much. I might have had an ulterior motive for buying it." I run my tongue over my lip and picture her on her knees. Basically, the image she just scarred her brother with.

"Oh yeah? What was your motive?"

"You're so fucking pretty, princess."

"Easton . . ." she protests, and I wonder if she even knows how pretty she is.

"You're mine, Madeline Hayes. And I want the whole world, including you, to know it. I thought a great way to start would be with my name on your back. And I may have a fantasy or two that start with you in my jersey."

"Oh yeah . . . ? Where . . . where do they stop, E?" Lindy's eyes flash wide with heat, and I know I hit the mark.

"You gonna watch the game tonight?"

She nods her head but doesn't say anything.

Hopefully, too busy thinking about what I just said.

"I'll call you afterward and tell you how the fantasy goes."

She takes a minute to collect herself, then slides her mask in place. The one she uses when she's on display for the world. The one I hate. "Have a good game, hockey boy."

"I'll talk to you after we win, wife."

LINDY

"Are you still hiding from your family?" Everly asks as she looks up from the sketch she's working on in our living room, then laughs when I don't answer. "Okay, so we're just going to avoid the Kingston conversation? Because I saw Scarlet today at cheerleading practice, and she asked me if I knew whether you met with an attorney yet."

"Of course she did. Because that makes more sense than picking up the phone and calling me." They drive me nuts. "I'm not actively avoiding them. I just haven't gone out of my way to see any of them."

Everly's eyebrows shoot straight up to her forehead, and yes, I hear how bad that sounded.

"Ma was asking about you when I saw her at the bakery earlier too." Maddox walks out of our kitchen and grabs the TV remote off the mantel. "Wanted to know if you figured out what you were doing yet."

"Don't you have your own place?" Everly kicks his leg when he sits next to her.

"Yeah, but Callen's fucking some cheerleader in there. Didn't feel like listening to the wannabe porn star, and I wanted to watch the game. See if Jace kicks the shit out of Easton on the ice tonight."

"Oh . . . which cheerleader?" Everly's eyes sparkle with excitement. "We sign a contract that we won't screw the players."

"How the fuck should I know? Want me to go interrupt them and ask for ID?"

Everly steals the beer out of Maddox's hand and takes a sip. "Oops. Backwash. Guess you need to get your own now, buddy."

"That *was* my own, demon spawn." He finds the Revolution game about to start and pauses the TV, then heads back to the kitchen. "You want one while I'm in here, trouble?"

"No," I call back and sit down on the other side of Everly. I peer over her shoulder at the stunning sketch of a ball gown with splashes of pink. "That's gorgeous."

She smiles nervously. Definitely not her norm. This girl was born confident. "I'm glad you like it because I have a favor to ask."

"Sure. What's up?"

"Let me design your dress for the New Year's Eve gala."

"Okay."

"That's it?" she questions, shocked. "Lindy, do you even realize how many people photograph you that night? Seriously, I think you were on the cover of *People* magazine last year."

"Okay. So I'll make sure to tell everyone I'm wearing an Everly Sinclair original." I pull my legs up on the couch as she tackles me.

"Oh my God. I love you."

"Dude, girl-on-girl is hot, but not when it's you two," Maddox grumbles and presses play on the TV.

"Whatever." Everly untangles us and fixes her hair. "Like you'd ever have a chance with us."

"Uhm . . . hello . . . ? Shared bloodlines here," I interrupt their argument, and they both turn to look at me.

"You got a new number on your jersey there, trouble." Maddox nods toward me, then hands me a bottle of water.

"You do," Everly inhales, overly excited.

"Calm down. It doesn't mean anything," I tell them both, and okay, so maybe I'm trying to convince myself a little too.

"She doth protest too much." Evie smiles.

"What the fuck did you just say?" Maddox asks, and I giggle.

Everly rolls her eyes. "How did you graduate from college?"

"I fucked a lot of TAs," he tells her matter-of-factly.

They don't stop, but I tune them out because there on my screen is Easton Hayes in my team's uniform, skating onto the ice. And oh my, does he look good.

"Oh, honey, I hope you climb that man like the tree he is, once he gets home. Look at him." I blush at Everly's words because yes, I'd very much like to climb that tree.

"Are you climbing trees now, trouble?"

I reach across Everly with a pillow and smack Maddox in the face. His beer spills down his shirt, but *oh well*. "It's none of your business whether I'm climbing trees, madman."

He snatches the pillow out of my hand and uses it to wipe his shirt. "You're right. I don't wanna know. But make sure Hayes knows if he hurts you, I'll kill him." All the joking is gone from Maddox's voice, and I'm pretty sure he's completely serious.

"Aww," Everly singsongs. "You do love us."

He smacks her with the pillow. "Nope. Not you. Just her."

THE KEEPER

LINDY

You won!

EASTON

Were you watching me, princess?

LINDY

I do own the team, hockey boy. I was watching everyone.

EASTON

Admit it. Not like you were watching me.

LINDY

Maybe.

EASTON

We're pulling up to the hotel now. I'll be in my room in a few minutes. Are you still in my jersey?

LINDY

Maybe.

EASTON

Leave it on.

LINDY

Aren't you sharing a room?

EASTON

Boone's going to the bar. He already asked me if I wanted to go.

LINDY

You should go and make friends.

EASTON

I'd rather talk to my wife. In my jersey. And princess . . .

Lock your door.

Holy hotness . . . okay.

I drop my phone down on my nightstand, and decide to lock my door, just to be on the safe side. I mean, it's not like we can do anything over the phone, right?

Can we?

Of course we can. Crap.

What's he thinking?

I've never even had real sex.

How am I supposed to know how to have phone sex?

I turn off my overhead light and fan, then decide to turn the fan back on. All my years on the ice have trained me to be more comfortable cold than hot. When I sit back down on my bed, I decide to turn on the lamp on my nightstand. It gives off a soft glow, and I wonder if I should shut it off, but before I can decide, Easton's calling me.

His gorgeous face flashes on my screen, and I take a deep breath and answer, "Hey."

His sandy-brown hair is still damp from the shower I'm sure he took after the game. A navy-blue Revolution hoodie is stretching across his chest, and a sexy smile sits on that handsome face. "Hey, princess. How was your day?"

I take a deep breath and try to convince myself I can do this.

It's Easton, and we're just talking like we've done a million times.

Only this time, I'm wearing his jersey and his rings and thinking about very un-friend-like things. I guess that explains why I'm suddenly so nervous, my hands are shaking.

"Well I avoided my mother. That was fun." I move my pillows around behind myself and get comfortable. "She called twice, then texted, asking me to meet her at Sweet Temptations for coffee tomorrow. So I've got that to look forward to. I only had two baby skaters cry through their lessons, which is one less than yesterday, so that's winning. I officially congratulated my former partner on replacing me. And I watched the Revolution game tonight with Everly and Maddox. Oh, and Everly asked me to wear one of her designs to the New Year's Eve gala. So there's that."

He drops the phone down, so I'm looking up at him as he takes off his hoodie. His gray t-shirt rides up with it, momentarily exposing his delicious abs before he tugs it back down and sits down. "That's a lot to unpack. Are you going to meet your mom tomorrow?"

"I don't really want to."

Easton shrugs. "Maybe she'll surprise you. I talked to Jules today, and she was pretty good about everything until she heard Jace. Then she went kind of crazy before I ended the call. Jace better be careful. You know how Jules can get."

"I do." I snicker. "Big brother better watch himself."

"What's this gala thing?"

I fill him in on the event my brother Max's wife started a few years ago as one of the many fundraisers she holds throughout the years to benefit the kids in our community. Then I listen as he tells me all about the game and my jerkoff brother, getting more comfortable the more we fall back into our old routine.

Eventually I reach over and turn off my lamp and get under my blanket.

"You tired, princess?"

"Yeah." I yawn. "Aren't you?"

"I am, baby." His sexy voice grows gravelly and pulls at something deep within me.

"I wish you were here," I whisper.

"Me too. Go to sleep, Lindy. I'll call you tomorrow."

"Stay with me until I fall asleep?" I ask quietly.

"Always." He adjusts himself until he's lying down too,

and I close my eyes like I've done a hundred times before with this man. "Sleep," he says softly, and I slowly drift off.

GRACIE

I've decided decaffeinated coffee is pointless. Kinda like a hooker who only wants to snuggle.

KENZIE

Have you ever snuggled a hooker or did you mean to text that to Callen?

EVERLY

Seriously. WTF is the point of decaf? Give me the strong stuff or give me nothing. Wasn't that a line in a movie?

BRYNLEE

OMG. NO.

LINDY

Well I'm about to walk into Sweet Temptations. Maybe I'll see if throwing decaf at someone is as effective as full caf.

BRYN

You can't throw coffee at your mom, Linds.

LINDY

We'll see about that.

I pull up in front of Sweet Temptations, my sister Amelia's shop. Once I'm out of my car, I adjust my coat and turn to look at Charles, who tends to give me a little more breathing room when we're in Kroydon Hills. "I'm pretty sure I'm safe in Amelia's shop. Why don't you go get lunch across the street?"

"You know I'm not going to do that, Miss Kingston. I'll be in the back corner. You won't even know I'm here," he tells me with a look of frustration growing on his face.

I shake my head and push through the pink doors. The sugary scent of freshly baked sweets mixes with the spicy smell of coffee and wafts through the shop, making my stomach growl. When I was a little girl, this was my favorite place to go. Mom would pick me up from ballet class at the twins' mom's studio next door and bring me here to pick out a cupcake and get a hot chocolate. I always felt so cool because Mom would have her coffee and I'd have my cocoa. That feels like so long ago.

The shop is mostly empty today. I guess the morning rush has died down already because Amelia is sitting at one of her mismatched, *Friends*-inspired tables, sipping a cup of coffee with Lenny and Scarlet. *Great*. I didn't realize this was going to be another full-blown sister thing. At least my brothers aren't here too.

I walk up to the counter and order a cup of tea and a scone from the woman behind the register, then join my sisters. "Mom's not here yet?" I ask as I slide into my favorite purple, crushed-velvet chair.

Lenny opens her mouth to answer as the bells chime over the door, then nods. "Jules and she just walked in."

"Great."

Mom and Jules co-own an event-planning company with offices across the street. I guess they came right from the office. Mom joins us at the table as Juliette walks over to the counter and grabs two coffees.

"Thanks for coming, honey. I hoped we'd have been able

to talk before now, but you don't seem to want to talk to me, so I thought maybe your sisters could help."

"Ganging up on me isn't going to help, Mom." I tear a piece of my scone and stuff it in my mouth before I can say anything else.

"We're not here to gang up on you, Lindy. But we need you to talk to us," Lenny tells me like she's rehearsed the words.

"Have you spoken to a lawyer yet?" Scarlet asks, and Juliette looks at her like she has five heads.

"Becket's her lawyer," Jules snaps.

Becks is a senator. I'm sure he has better things to worry about than his little sister getting drunk and getting married," Scarlet responds, and I nearly get up and leave right then and there. But the look Jules gives me stops me in my tracks.

"There is nothing more important to Becket than his family, Scarlet. Easton and Lindy are *both* his family. Don't forget this man you all want to vilify for marrying Lindy isn't a villain. You know him, and you love him. Try to keep that in mind." She throws me a quick look which holds a hint of defiance, then turns back to the table. "Becket's already spoken to Easton."

"Good for Becket." Mom looks so hurt, I almost feel bad. *Almost*. "Why didn't you tell me earlier? Is he filing for the annulment?"

Jules looks at me again but answers my mother. "Not yet. Easton and Lindy are both adults. If they want an annulment, they know who to talk to. But neither of them has asked for one yet, and I have to think there's a reason for that."

"What kind of reason could there be for not annulling a quickie Vegas wedding? She was clearly drunk in the pictures," Scarlet demands. "Nothing good happens in Vegas."

"You got pregnant with Killian in Vegas, if I remember correctly, Scarlet," Lenny jumps in.

My head snaps to Lenny.

Do I have a sister on my side?

"I'm not saying my son was a mistake, Eleanor. But you've got to remember how difficult those next few months were. Everyone was furious."

"And by the end of the summer, you were in love, married, and about to give birth. You were happy, Scar. Sometimes things happen because they're supposed to." Lenny reaches over and squeezes my hand.

Holy hell. I've got an ally.

"I was an adult with a career and life experience on my side," Scarlet bites back, and my mom nods her head.

"Madeline, honey. You're still so young and so sheltered. That's my fault, but it doesn't change the fact that we've never forced you to be an adult." I can't look at my mom by the time she's finished speaking.

Anger fuels me when I can finally speak. "By the time you were my age, you were a widow with a baby. Lenny was living with Bash at twenty-three. Amelia was marrying Sam." I look around the table, and Jules shakes her head.

"Don't look at me. I was living my life, working my way through the hottest runways with the most in-demand designers in the world at twenty-three. Your brother came into my life when he was supposed to. When we were ready for each other." She turns and squeezes Mom's hand. "I think you need to let Lindy and Easton decide what they're ready for, Ashlyn."

"You don't get to tell me what I need to do with my daughter," Mom snaps back with an icy tone I've rarely heard her use and never with Jules.

"I need everyone to stop and listen to me." I'm proud of how steady I manage to keep my voice, even though I feel like I could scream or cry at any moment. "I've let you all treat me like the baby my entire life. And I get it. To you, I'll always be the baby of the family. Even though I have nieces and nephews nearly my age and others who might as well still be in diapers, they're so much younger. But for some reason, it made you all feel good to baby me, and I let you."

"Madeline—" Mom interrupts, but I cut her off.

"No, Mom. I can't even go anywhere without a bodyguard. Does anyone else have that?"

"I do," Scarlet says. "And so does Max."

"You're both the faces of the company," I argue.

"Caitlin and I both have bodyguards, Madeline. But I'm aware we're a little different," Amelia says softly. "And you are high-profile. You have to be more careful than most of us. People are crazy. We all know that, you especially."

Amelia's husband, Sam, owns the security company my bodyguards all work for. I get that she'd know better than most what I need, but I just don't care. I'm past that point. "I appreciate that. I really do. But I'm done asking for you all to treat me like an adult who can make her own decisions and live her own life. Do what you want. But I'm no longer asking. I shouldn't have to." I stand up and look at Charles, who's tucked into the corner of the room. "Charles."

His head pops up, and he attempts to act like he hasn't been able to hear every word we've said.

"You're fired."

"Sorry, Miss Kingston. But I don't work for you."

"You don't," I agree. "But if you follow me out of this shop, I'm going to call the police and tell them you're harassing me." I slap my hand against the table and startle everyone. "None of you have asked me. Not one of you. I said this a few days ago, and you still haven't asked me."

I shove the big chair out of my way and grab my purse.

"Maybe if one of you had bothered to ask me what I wanted, this would all have gone differently. But you didn't. So here's how it's going to go. I have my own money, my own job, and my own condo. I own my own stock in King Corp. John Kingston might not have ever been a father to me before he died, but he set me up so I can lead an independent life without ever having to worry about how I support myself. If I want to teach baby skaters forever, I can. I don't have to ask you for anything. I never have. So you no longer get a say in my life."

I look around at my mom and sisters and stop on Lenny. "Except for Len. She at least gave me the benefit of the doubt. Maybe you should all try that."

"Madeline . . ." Mom's voice shakes, but I'm past the point of caring.

"Do. Not. *Madeline* me. I love you. But until you realize just how wrong you are, I'm done."

And with that, I storm out of Sweet Temptations and into my car. Charles doesn't follow me, thankfully, and I manage to force myself to keep my shit together long enough to drive home. It's not until I'm parked in our garage that I finally let the tears fall.

My phone rings a few minutes later, and Easton's name appears, so I try to pull it together. I look in the mirror and wipe my face, trying to clean myself up before I finally answer. "Hey," I sniff.

"You okay?" he asks.

"Of course." My voice cracks on the lie. "Why wouldn't I be?"

"Jules called. She said you fired Charles and told your mom and sisters to fuck off." His lips tip up in a hesitant, slightly crooked grin. "I would have paid to see Scarlet's face."

"Oh, she was pissed," I admit. "If I wasn't so angry, I might have enjoyed it. I'm so mad I let it go this far. It's as much my own fault as it is theirs."

I look down at my phone, nervous Easton's going to be one more person who's mad at me after today. "Are you going to yell at me for firing Charles?"

"Baby, you're a grown woman. You're smart and beautiful and the furthest thing from irresponsible there is. If you wanted to fire him, you had every right to. This is your life. Just do me a favor and try to be aware of your surroundings until I get home. I don't want anything to happen to you."

"Easton . . ." I trail off, blown away by his trust in me. I stare at him, wishing he were here for the second time in twenty-four hours.

"I've got to go, princess. We're boarding the plane to head to Washington."

"Okay. Enjoy your night off and kick ass during your game tomorrow." I end the call, wondering how this man—my husband—can have such blind faith in me, but my own mother can't.

I get out of my car and run into Gracie. Her hair is pulled back in a bun, and a heavy puffer jacket is thrown on over her sweats and Uggs. "On your way to practice?"

"Yeah. I just let Myrtle out. She's sleeping." Grace points at my face. "What's with that?"

"With what?"

"That look on your face. You're up to something."

Am I up to something?

I guess maybe I am.

"I think I may be falling for my husband," I tell her as an idea comes together.

"Yes. I love it. I'm here for all the romance. And you and Easton will make pretty babies."

"Calm down, good twin. Let me figure out this whole marriage thing before we start talking babies."

Gracie laughs as she walks by me and opens her car door. "Just saying, I call godmother."

She shuts the door, and I smile as I decide I'm here for the romance too.

I shoot Bryn a quick text.

LINDY
What hotel are you guys staying in tonight?

BRYNLEE
Why?

LINDY
I need to know where I'm going once I land.

The Philly Press

KROYDON KRONICLES

A ROYAL RUMBLE

Sources close to the Kingston family tell this reporter there's trouble within the royals. The Kingstons aren't happy their baby has married the hockey hottie and staged an intervention at everyone's favorite sweet shop earlier today. Will this be the last straw for our lovebirds or will America's sweetheart stand by her man? Only time will tell.
#KroydonKronicles

EASTON

"Wanna grab dinner?"

I look over at Boone, who hasn't stopped staring at his phone. Over the past twenty-four hours, I've learned he's not just our co-captain. As far as I can tell, he's also our social director. He may have actually missed his calling. "Is Kingston gonna be there?"

"Come on, cradle robber, you know he is," he taunts with a goofy-ass grin.

He also never seems to take anything but his game seriously.

On the ice, he's a killer. Off the ice, he's the joker.

"Fucker. She's barely five years younger than me."

"Yeah, I know. But it's so much fun to piss you off. Kingston's sister is like this on-off switch for you. Him too, now that I think about it."

"Yeah. I'm gonna pass on dinner." I'm not willingly forcing myself to deal with Jace any more than I have to.

"You're never gonna mesh with the team if you keep hiding in here, Hayes." He's not wrong, but there's not a chance in hell I'm telling him that or that I'm having dinner with Jace.

"Listen, I'm trying to respect Kingston and steer clear. At least for now."

"We got three more days, Hayes. You gonna hide the whole time?"

There's a knock on the door, and Boone waits to see if I'm gonna change my mind.

I'm not.

"All right. But I'm gonna get you to come out with us at some point," he tells me before he grabs his coat and leaves.

Fuck.

Not two minutes later, there's another fucking knock.

"Dude. What? Did you forget your key card?" I yank the door open and stop, frozen in place. "You're not Boone."

"No," Lindy's eyes light up. "I'm not."

"Princess . . . how are you here?"

She lifts up on her toes and kisses my cheek, and I swear I feel that one fucking touch everywhere. "Wanna run away with me?"

"What?" I laugh. "I kinda signed a contract that says I need to be here tomorrow."

"That's okay. It's just one date. I'll have you back tonight. You game?"

"To run away with my wife? Fuck yeah, I'm game." I grab my hoodie and take her hand in mine. "Where to, Mrs. Hayes?"

She stops and looks at me, her eyes fucking shimmering like I haven't seen in a long damn time. "I might not hate the sound of that as much as I used to."

"I guess that's a start." I squeeze her hand. "Want to grab something to eat?"

"Uh, uh, uhh. I decided it was my turn to do the wooing.

I've got plans for us." She pulls a box from behind her back and hands it to me.

"What's this?" I pull on the black-velvet bow as Lindy smiles at me.

"Open it and see."

I lift the lid and the tissue paper and find a light-gray peacoat and a black cashmere scarf. "You bought me a coat?"

"Yeah, hockey boy. It's snowing in Washington, and you've only got a hoodie with you. Now put it on, and let's go."

I slide it over my arms, and Lindy takes the scarf and folds it around my neck. "So handsome." She smiles at me, and the world feels fucking right. "You ready?"

"Lead the way." I follow my wife down through the lobby, where a town car is waiting for us out front.

The driver moves to open the door, but I cut him off and hold it open for Lindy, then slide in next to her. "You know, Jules warned me about you Kingstons and your wooing."

"Can we please not talk about my family tonight? I need a little distance." She crosses her legs and folds her hands in her lap nervously. "Tonight, I'm embracing being a Hayes."

"Princess . . . I love you being a Hayes. But a few days ago, you weren't even sure you wanted to be married. You can't run away from your family forever. And when you decide you're ready to be mine, I'm gonna need it to be because you want me, not because you don't want them."

"I know." She lays her hand over mine. "But I realized something when you called earlier."

"Oh yeah? What did you realize?" I ask her and bring her knuckles to my lips.

"I realized standing up for myself today wasn't me running away from my family. It was me running toward my life. And I want that life to include you."

"Lindy . . ."

The driver stops the car. "We're here, Ms. Kingston."

"Thank you. I think we should be about an hour."

I look through the tinted windows at the twinkling lights in front of us. "Where are we?"

"You'll see." She opens the door and tugs me out after her. "I just thought we could use a little fun tonight."

We walk through a roped-off parking area to a small ticket booth and then into a Christmas Village. Holiday lights are strung across the aisles, highlighting booths full of food and games. A twenty-foot-tall, lit tree is off to one side, with Santa sitting in a big red, regal-looking chair and a line of kids in front of him. In the center of it all is an outdoor ice rink and skate rental. I tug down Lindy's soft white hat until the fuzzy white pompom bounces. "We going skating, princess?"

"We sure are, hockey boy." She tugs on the collar of my peacoat. "Wanna watch me kick your ass?"

I wrap my arm around her waist and squeeze her ass. "I'd much rather *watch* your ass than kick it, baby. But if you think you can skate faster than me, I'll race you. Just don't think I'm letting you win."

"Bring it, hockey boy."

I spend the next hour looking for reasons to touch my wife.

To hold her hand.

To grip her hips. Her waist. Her face.

No one bothers us. Hell, no one even realizes who we are until we're sitting on a bench once we're done and taking our skates off. A little girl with a purple hat and matching mittens stops in front of us with a napkin and pen held in front of her. Her mother stands off to the side, silently watching her. "Excuse me. Are you Madeline Kingston?"

"That depends." Lindy smiles. "Do you promise not to tell anyone?"

The girl pulls her mitten off and holds up her pinkie finger. "Pinky swear."

Lindy pulls off her glove and links their pinkies. "Then, yes, I'm Madeline Kingston. Do you like figure skating?"

The little girls eyes grow as big as saucers. "I'm Sarah, and I watched you win the gold medal last year. You were amazing," she says with this level of awe in her voice that makes me want to say, *Yeah, kid, she really is that great.* "I want to skate like you one day." She looks over at me and tilts her head. "Are you her partner?"

Lindy rests her hand on my leg. "Can I tell you a secret?"

Sarah moves in closer and nods her head excitedly.

"He's better than my partner. He's my husband."

Fuck . . . what those words do to me.

"He plays hockey," Lindy tells her.

"Hockey?" Sarah's face pinches. "Ehh. I like figure skating better than hockey."

I think I just got dissed by a first grader.

"Would you sign this for me?" Sarah shoves the pen and napkin toward Lindy, who does the sexiest thing I've ever seen. She signs the napkin.

XO,
Madeline Kingston Hayes

Lindy

Once Sarah and her mom walk away, I turn toward Easton to ask if he wants to get something to eat, but the heat in his eyes stops me. "You okay, hockey boy?"

"You planning on taking my name, princess?" His voice is thick with emotion, and suddenly, it's just him and me. The rest of the rink fades away, and something tugs at the back of my mind. Something *important*.

I move onto his lap and cup his face in my hands the way he seems to like doing to me.

"I was thinking about it," I whisper against his lips, giving him a truth I'm not sure I even realized until now. "Is that okay with you?"

"It's your choice, baby. You hold all the power. You always will." Easton's hand grips my head as he deepens our kiss, and I get a sense of déjà vu. His tongue licks into my mouth, and I hum a quiet moan. "You ready to get out of here?"

"Yes," I whisper breathlessly, and Easton stands with me still in his arms.

"Put me down, hockey boy. We can't scare the kids."

"Fuck the kids." He kisses me again, and I almost agree. *Almost.*

"Easton..." I pull back.

"Fine." He drops my legs and lowers me to the ground. "Tell me you got a room at the hotel, because I'm sharing with Boone, and I don't want an audience for what I'm about to do to you, princess."

"Better. I got a suite."

Easton groans and takes my hand in his. "Let's get the fuck out of here."

EASTON

*L*indy rests her head on my shoulder and her hand on my heart for the short drive back to the hotel. Neither of us speak as the air around us grows thick with need and want and fucking desperation.

Her fingers resting under my coat and over my shirt somehow still scorch my skin. Meanwhile, I don't dare move a fucking muscle. I can't. Not yet. Because when I finally do, I'm not sure I'll be able to stop.

Lindy looks up at me once we get back to the hotel, and I run my hand over her hair and hold the back of her head in my hands, framing her face. Those long lashes flutter around her stormy-blue eyes, dragging me under like a riptide.

Slowly, she drags her teeth over her trembling lower lip and leans into my hand, and I finally allow myself to give in to the desperation coursing through my veins. With the smallest fucking move, I lean down and ghost my lips over hers and swallow the sweetest sigh as it falls from her lips.

"Take me upstairs, Easton." She pulls back and runs her fingers along my lips. "Please..."

This woman... "As you wish, princess."

Slowly, we get out of the car and make our way through the lobby to the bank of elevators. She pushes the button for the fortieth floor, and we stand, silently holding hands while we wait. The heat between us threatens to incinerate anyone in its way. A few teammates walk by and say hi, but I just give them a chin lift in return as Lindy's lips press into a slow, sexy smile.

We step onto the elevator with a handful of other people, and I move us to the back corner and drag her in front of me with her back only centimeters from my chest. Not touching. Not in an elevator full of people. Not with the paparazzi staked out all over the hotel.

We both watch the numbers tick by at a rate a fucking snail could outpace.

She steps back and links a single finger with one of mine, and for such a small movement, it's one of the hottest experiences of my life.

Because it's her.

Because I want her more than anything I've ever wanted in my life.

More than my own life.

Just her.

One of my teammates gets on at the thirty-fifth floor and spots me. "Hey, man."

I lift my chin and nod at Donnelly.

He looks from me to Lindy, then down at our hands, and the doors to the elevator close again. "A few of us are hanging in Malcom's room. He always brings his gaming setup with him for the long stretches. You want in?"

"Not tonight, man. But thanks," I tell him, my voice sounding strained to my own ears, like I swallowed broken fucking glass.

His eyes drop down to our linked fingers, and Lindy

drops my hand. I'm about to protest when she reaches out for him. "Hi, I'm Lindy, Easton's wife. It's nice to meet you."

He shakes her hand, then looks at me, surprised. "Damn, Hayes." I get ready for him to give me shit, but the asshole laughs instead. "You out-kicked your coverage, man."

I grab her hip just before the doors finally fucking open on our floor. "You have no idea. See you tomorrow."

I guide Lindy out of the elevator and hear Donnelly's low whistle as the doors close.

Lindy pulls her keycard out of her purse with shaky hands and struggles to unlock the door. So I take the card from her and hold it against the sensor until it blinks green and push us inside.

Once it closes behind us, I spin us around and cage her in, leaning a hand against the door and towering over this beautiful woman.

Her breath shakes as she leans her back against the door and grips the collar of my coat. "Easton . . ."

"Your speed, baby." I pull the white hat off her head and tug on a lock of her long hair. "You're in control, Lindy. This is your choice." I run my hand along her throat and cup her chin. "You decide how tonight goes."

She shrugs off her own coat and lets it fall to the floor, then pushes my coat off my shoulders and wraps her arms around my neck. "I choose you, Easton." She licks her lips and smiles. "I choose us."

And that's it.

The strings holding me back—the ones that have been slowly fraying—snap.

I lift Lindy from her feet and crash my mouth over hers as she molds her body to mine. Her arms and legs wrap around me, begging to be held. Like I'd ever let her go.

I take two steps toward the king bed in the center of the room and lay her down in front of me. The moonlight filtering in from the hotel window illuminates her creamy skin. "You are so fucking pretty, princess."

Her fingers move to the buttons on her silky shirt, and she slowly pops each one open until she sits up on her knees and shrugs her shirt off. A lacy white bra cups her perfect fucking tits, and my mouth waters as she reaches for me.

"I trust you, Easton." Her cold hands reach under my shirt and press flat against my skin. "Like I've never trusted anyone else in my life. Make me yours."

I swallow the emotion clogging my throat and rip my shirt over my head, then shove out of my jeans and yank hers down her legs. I need to see her. To feel her skin against mine. I drop a knee on the bed and wrap a hand around her throat. "You've been mine since the day I chose your life over mine, baby. All mine. Always mine. Even if you didn't know it yet."

Lindy

*H*is words are like a balm to all the tattered edges of my soul.

He kisses away the tear I hadn't realized had fallen from my eye. "You are the most gorgeous woman I've ever seen, Lindy." His tongue invades my mouth as his hands skim over my skin. A callused palm cups my breast, and my nipple aches from the feel of his thumb brushing over it.

"Easton . . ." I plead and wrap my legs around his hips, dying to feel him against me. Inside me. Only his boxers and my panties separate us, but it's too much. My fingers reach for his boxers. "Please . . ."

"I know what you need, baby," he teases and bites down on my bottom lip, sending a shockwave straight to my core.

I lift my hips and grind against him, desperate for relief, and his hands quickly grip my hips and press them to the bed.

Big fingers skim the waistband of my lace panties as his hazel eyes deepen to a dark green right in front of me. "I've fucking dreamed of this moment for years, wife. I'd do anything for you, but even you can't convince me to rush it."

His tongue trails along my neck and down to my breast as he sucks my tight nipple into his mouth through the lace, and I cry out.

One hand slides from my hip into my panties, and goosebumps break out over my already-heated skin before his fingers drag through my sex, sending me spiraling through the stratosphere. "Your pussy is soaked, baby."

Those green eyes look up at me for a moment, and that dimple I love pops deep in his cheek as he slips one blunt finger inside me. Then another. Stretching me. Filling me in a way no man ever has. His rough thumb rubs circles around my clit, and I moan, frantic and needy.

"Is this for me, Lindy?" he groans as he licks into my mouth. "Are you wet for me, baby? Have you been dreaming about me fucking you?"

"Yes," I cry out as his filthy words make my clit throb with need.

My stomach quivers when nerves mix with excitement. "Give in, Lindy. Give me what I want, and I'll give you everything you need."

His rough thumb circles my clit, teasing me. Torturing me. Working me into a frenzy. "Easton," I beg as his fingers curve to hit a spot deep inside me.

My core tightens as Easton looks down on me with a heavy-lidded heat in his dark eyes.

He pulls his fingers out, and my eyes fly to his, suddenly empty and confused.

Desperate for this not to end.

But he just smiles that lazy smile and traces my lips with his fingers, pushing them onto my tongue. "Suck."

I tentatively dart my tongue out and trace the pads of his fingers, then do as I'm told and suck his fingers clean. My own tart taste explodes against my tongue, and Easton's already endlessly dark-green eyes deepen.

He pulls back as I swirl my tongue around his fingers. "Good girl."

Pleasure blooms deep in my chest, liking the sound of that. I lean up and capture his lips with mine, sharing the taste. Pushing my tongue in his mouth and letting him suck it.

I scrape my nails down his chest and around his waist. Digging into the soft skin above his ass, I moan. "Easton..."

"I know what you need, baby." His hands find the sides of my panties and tear them from my body.

Fuck, that's hot.

He moves off the bed and drops to his knees before he yanks me to the edge, throws my legs over his shoulders, and settles himself between my thighs. "I've been waiting a lifetime to taste this pretty pussy, princess."

Dark eyes peer up at me with a wicked grin, and I wonder how I ever thought I could live without him.

"That's my good girl. You doing okay, baby?"

I nod my head, unable to answer him as emotion overwhelms me.

"I need your words, wife. Remember. This is your show. I need to know you're okay."

"Don't stop, husband," I whisper and lean back on my elbows, unable to look away from the fucking god of a man between my legs.

"Fuck, baby. Never gonna get tired of hearing that word on your lips."

His words fan the flame of my already-insane desperation.

Easton drags his flattened tongue slowly through my pussy.

His eyes stay locked on mine as he laps at me. Then he growls against my sex.

Vibrations pulse through me, and I pant, unable to catch my breath as I watch the sinful sight in front of me. I dig my fingers into his sandy-brown hair and tug as I press myself against him.

He spears my pussy with his tongue, then slides a finger in so he can go back to tonguing my clit. And, *oh God*, it feels so good. Too good.

He pushes down on my pelvis with a flat palm and feasts on me until I'm screaming. Sucking my clit. Teasing me. Bringing me to the brink. Until darkness tinges the corner of my sight, threatening to overwhelm me, and my entire body is lighting up. Electricity covers every inch of my skin.

I tug his hair, unable to think or speak or focus on anything except the way his mouth feels against me. And just when I think I can't take it anymore . . . when my body refuses to believe anything can ever feel better, his teeth finally scrape over my pulsing clit, and my back arches off the bed, while I scream out his name, and explode on his tongue.

Easton drags his tongue through my drenched sex and over my clit one more time, absorbing the aftershocks wrecking my body. Then he grabs his wallet from his jeans and smiles devilishly. "Such a good girl."

He cradles my face in his hands and drags his mouth over mine.

Kissing me reverently.

"You okay, baby?" he whispers against my lips, and I nod my head and notice the gold condom packet held between his two fingers.

I take it from his fingers and toss it to the floor. "I'm better than good."

"Princess . . . It's my job to keep you safe. We can't just toss away the condom and play pull-out and pray."

"I've been on the shot for years, E. The pill made me gain weight I couldn't have on if I was going to continue to be thrown through the air. I started getting the shot at seventeen." I hold his eyes with mine and melt under his touch. "You love telling me you're my husband. Show me what that feels like. Nothing between us."

"Don't you want to talk about STDs or previous partners?" he presses.

I shake my head. "You're the only man I've ever been with, Easton. And I trust you with my life." I run a thumb over his cheekbone and get lost in his eyes. "That's enough for me."

"I'd never do anything to hurt you, Lindy." He rolls me to my back and hovers over my body.

"I know."

I reach down and press Easton's cock against my entrance, then look up at him, not sure he's going to fit.

"It'll fit, wife." A muscle ticks in his jaw, and I brace my hands against his chest.

"Fuck me, husband."

"As you wish," he whispers against my lips and pushes himself inside me.

Pleasure spikes, stained with pain as he stretches me so thoroughly, my body vibrates around his, and I tighten my knees against his hips, worried he's going to split me in two.

"Breathe, baby." He pulls back out before pushing back in further.

Tiny movements, achingly slow. Working his way in, so he doesn't hurt me. "You gotta breathe, Lindy."

My eyes lock on his, and I'm suddenly struck by the weight of the moment.

I never thought I'd be here. With him.

He pulls back again. "Deep breath, princess."

And when he thrusts in this time, the pain sears and splinters, and I gasp, searching for breath.

Easton presses his lips to mine, firm and soft and so utterly hypnotic. I get lost in the perfection of it all for a moment. He pushes his tongue into my mouth, and his fingers play with my clit.

I cling to my husband, slowly adjusting to his size. And after a few moments, the pain falls to the background, and it's just us and the slowly building pleasure.

I slide my hips against his, trying it out.

Seeing what I like.

Wanting this to last forever but needing more.

Chasing the high I crave that only Easton can give me.

"You're a fucking vision, wife." He rolls us over so I'm straddling him and grips my chin. "Ride me, Lindy. Show me what you like."

Pleasure pounds through me as I press my palms flat against his chest and try to move. My head falls back, my hair brushing against his thighs.

"That's it, baby. You take my cock like such a good girl."

One hand slides to the center of my back, and I tremble beneath his touch as we find our rhythm.

"Your fucking cunt was made for me, baby."

My entire body clenches at his filthy words.

Wanting more.

"You like that?" he asks against my lips as his hips pick up speed, and I feel my arousal soaking our skin with each new thrust.

I practically purr as sensations overload my body, completely unable to speak or think or form words. I just cling to him.

"You like the way my cock fucks you, baby? The way you fit around me? Only ever me, wife."

"Only ever you," I echo back and claw at his strong shoulders as he fucks me harder.

Sends me higher.

My orgasm teeters just out of reach until Easton slams into me, and flames lick up my skin in a fiery explosion. Vivid color bursts behind my eyes as my orgasm washes over me, and my husband shouts my name out into the quiet room, then holds my face in his hands and covers my mouth with his.

He holds me draped against his body.

I'm unable to move and unwilling to care.

I tuck my face into his neck and suck his skin. "I think if I died right now, I'd die happy."

Easton's entire body goes rigid, and he stills beneath me.

"I love you, Madeline Kingston," he whispers, a quiet prayer from his lips. "You're not allowed to die until we're old and gray and lying in bed together. And then you have to wait until I've been dead for at least ten minutes, so I never need to know a world without you in it."

I lift my head and stare at him in shock. "What?"

He looks at me, confused.

Something tickles the back of my mind, and I stare deep into the depths of his eyes.

I stare at the man I married.

At the man who saved my life.

The lifeline who's made sure he was only ever a phone call away.

Who's been here, beside me, for a lifetime.

"I promise to love you until we're old and gray and

surrounded by grandbabies who love to skate," I whisper back.

And my God, the smile that breaks out on his face is everything.

It's agony and ecstasy and relief. It's excitement and love.

So much love.

"Princess . . . You remember?"

I don't bother trying to hide my tears, as I whisper in awe, "I remember everything."

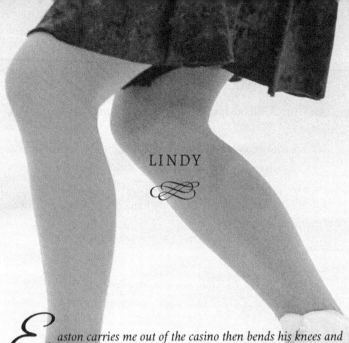

LINDY

Easton carries me out of the casino then bends his knees and slides me down the front of his body until my feet touch the ground outside. I slap his chest, then grab his arm to steady myself as the blood rapidly starts flowing back out of my head. "Listen, hockey boy. I don't like being upside down when I don't have to be."

"You didn't have to be upside down, princess. You chose to stay in that position," he bites back as he wraps an arm around my shoulder for support. And oh my, does his warm skin feel good against my bare back. His fingers play with the two hanging ribbons of the silk bow tying the top of my halter together until he starts letting them dance up and down my spine. Like I needed any more reason to be turned-on right now.

And okay, fine. I may have stayed in that position, dangling over his shoulder, because it put my face in front of his ass, giving me one hell of a great view. So sue me. It's my birthday. Consider this part of my present.

I mean, come on.

He's a fucking professional hockey goalie.

His buns are made of steel.

I wonder what else is made of steel.

"Where to, bitches?" Everly exclaims, then pushes us forward. "And wherever it is, let's go fast. Maddox is running interference with Charles to try and get us some unsupervised time. So move it, people."

A party bus sits on the side of the street with the engine running, and a bunch of drunk women with bachelorette sashes wrapped around themselves stand in front of it. All except the one I'm guessing is the bride, based on her white dress and tiara.

Pace laughs and grabs Easton. "Let's make you useful, E."

Easton tugs me behind him as Pace introduces him to the ladies and lets them know who he is, then offers them a thousand dollars to let us on their bus. And before I know what's happening, we've all piled on, and the driver is pulling away.

My girls all find seats with the bachelorettes as more champagne gets passed around and Easton pulls me down onto his lap.

"What do you think you're doing?" I ask with absolutely no intention of moving off his lap. Especially when I feel the lovely bulge between his legs.

I mean, I know I'm not sloppy drunk right now, but our earlier shots have absolutely given me just enough lady balls to lean back and enjoy the feel of my longtime crush beneath me.

"Conserving space, princess."

"Huh?"

"What I'm doing. I'm conserving space." He laughs and pulls on my hair.

I look at his hazel eyes and smile. "They're green today."

"What's green?" he questions.

"Your eyes. When you call me in the middle of the night, they're usually a caramelly brown with yellowish-green flecks. Almost golden. They're a mossy green when you're in a good mood during an interview after a game and brown when you're in a bad mood. They're a darker green today. What's that mean?"

His lips tick up in a crooked smile, and that dimple I love peeks through. "Guess you'll have to wait and see."

"Where are we going, ladies?" Callen asks, and when I look at him, he no longer has the two women from the club with him. Now two of the bachelorettes are fawning all over him. Good grief.

"Hey, man whore," I call out, and every single set of eyes on the bus turn my way.

Easton laughs and tucks my hair behind my ear, then leaves his hand anchored there as he whispers, "So . . . maybe you don't need to say everything that pops into your head."

"Whatever," I grumble, then look back over at Callen. "Does anybody know where Maddox is?"

"Chillax, birthday girl. Madman is waiting for a text from me. He'll meet us wherever we land," Callen announces before he drops his face back down into a pretty redhead's big boobs.

I mean, they're huge.

"You see them, right?" I ask Easton.

"See what?"

"Her boobs. I'd die for boobs that size," I grumble and look down at my own. "Maybe I should get a boob job."

"Don't you fucking dare," E snaps.

I turn around and enjoy the feel of Easton's other hand as it slips to my ass.

"First, why would you think you'd get a say in what I do to my body? And second—"

He moves his hand slightly so it's cupping a cheek. Like actually grabbing a big old handful of my ass, which, by the way, is significantly better than my tits. I have a great ass. "Wait . . . what was I saying?"

Easton's smile softens. "You were promising me you'd never change your body because it's fucking perfect exactly the way it is." I can practically taste the whiskey on his breath as he leans in closer, and the grip he has on the back of my neck tightens, sending uncontrollable goosebumps spreading out all over my skin. "You are fucking exquisite, princess."

I open my mouth for a snarky comeback, but there's nothing to say. Nobody has ever called me exquisite before. Instead, I settle for, "Your hand is on my ass, hockey boy."

"It is," he agrees. "Do you want me to move it?"

The bus pulls to a stop in front of a gorgeous casino with a pretty fountain light display, and everyone starts filing off but us.

Easton and I don't move.

Bryn and Kenz stop in front of us and stare, like we're animals at the zoo. "You guys coming?"

Easton shakes his head. "You okay if we catch up?"

Kenzie shrugs, and Brynlee looks at her phone, then back at me. "Yeah, we're good. Her tracker app is on. Text if you need anything."

Once they're off the bus, I stand up and stretch, then reach my hands out for Easton. "Come watch the lights with me. I love this display."

He takes my hand in his and walks me over to the bridge. "What was that about a tracker app?"

"We all have tracker apps on our phones, so we can find each other. They came in handy at parties during college."

E takes my phone out of my pocket, holds it up to my face to unlock it, then does something with it and then his before giving mine back to me.

And I just stand there, slightly confused, and let him.

"What the hell did you just do, Easton Hayes?"

"I added myself to your app and downloaded it on my phone," he says, like it's the most normal thing in the world.

"Why would you do that?" I ask, even though I kinda love the fact that he did.

Without skipping a beat, he cups my face in his hands and brings his lips to mine.

So close but not actually touching. "So I can always know you're safe."

I could blame the alcohol or the Las Vegas energy thrumming between us for what I do next. But either one would be a lie. In reality, I just want to do it.

I raise up on my toes, closing the distance between us, and press my lips to his, and . . .

Oh. My. Goodness.

They're perfect. Firm and soft and unbelievably delicious.

The electricity between us sparks and fires and lights up every nerve-ending like a Las Vegas light show.

Easton takes control, angling my head and deepening the kiss, and I think time and space cease to exist.

I run my hands through his hair as I press myself up against him, wanting to feel him, more of him, every inch of him. His erection presses against me, and I moan.

He licks into my mouth and, I swear to God, growls as he leans me back against the railing. Growls. Like his reaction is visceral. Primal.

"Easton," *I whisper against his lips, dragging myself out of this lust-filled haze.* "What . . . what are we doing?"

E takes a step back and looks at me for a charged fucking minute, then grabs my face again and crushes his lips to mine. "What I've wanted to do for-fucking-ever."

Good enough for me.

I wrap my arms around his broad shoulders and cling to him, like I'm scared if I let go, this daydream will end and reality will come crashing down on us like the water in that fountain.

"I'm so fucking tired of acting like I'm just your friend, princess." *He takes my mouth in a desperate kiss.*

"Tired of acting like you're not my girl." *One strong arm slides around the center of my back and presses us impossibly closer together.*

"Like I don't want you. Like I haven't always fucking wanted you."

Holy hotness wrapped up in a hockey-player package.

I'm pretty sure I just soaked my panties.

"Easton . . ." My hands claw at the front of his shirt, refusing to allow any space between us.

"Tell me you feel this too. Tell me it's not one-sided and I didn't just make a fool out of myself."

I run my teeth over my bottom lip and try to force the words to come out of my mouth, but nothing happens. I stand there in front of this fountain in Vegas with this beautiful man, shocked to my core. "Easton . . . I . . ."

Words escape me, so instead I tug the front of his shirt and pull his face down to mine.

He presses his lips to my forehead, and I slide my hands up the length of his neck to his face and hold him there.

"You gonna break my heart, princess?"

"God, Easton. Do you really not know?" I close my eyes and decide there's no going back now. "It's you. It's always been you, and it's going to always be you. I have loved you . . . God, I've loved you for as long as I can remember. I want to be old and gray and sitting on our porch one day a million years from now, surrounded by all our grandchildren who love to skate."

He swallows a deep breath and smiles against my skin. "Hockey skates or figure skates?"

"Mine are better. Sharper," I tease.

"Promise me something, baby,"

"Anything." I pull back and look deep into those green eyes that are growing darker by the minute.

"When we're old and gray, you've gotta let me die first, okay? Because I don't think I'd ever survive without knowing you were breathing on this Earth."

I run my fingers through the hair at the nape of his neck and tug. "How about I promise that when we're really old we'll die in bed together like that couple in the movie you used to hate. The one Kenzie and I use to watch all the time. I'll let you go first, but I'm gonna follow you, okay?"

"You sure, Lindy?" The look on his face makes me hesitate for a second.

"Marry me, Easton."

"What?" he chokes back.

"I'm standing here, in front of the man I've already loved for a lifetime, and he's promising to love me forever. There are twenty-four hour chapels all over this town. Marry me. Make me Mrs. Madeline Hayes." I giggle. I used to doodle that on my notebooks when I was a little girl.

"You're fucking serious?" He takes a step back, and I'm pretty sure my heart drops out of my chest. Please, God, don't let him laugh in my face. I don't think my heart could survive it. "Madeline Kingston, I'm supposed to ask you to marry me. It's supposed to be fucking romantic. Fit for a princess." His tone is pointed, but his smile is huge, and my heart soars instead of cracking.

"Fuck romance, Easton. We've already done life and death. We've already been each other's person for years. I don't need hearts and flowers and pretty words. I need you. And you say fuck a lot more than you say anything else." I bite down on my bottom lip, and he gets all growly again.

"You keep biting your lip that way, princess, and we're gonna need to hurry up and get married so I can put something else in that pretty mouth to keep you busy."

And there go my panties again.

"Okay. You find the chapel, and I'll get the girls." I step back, but Easton grabs my hand and pulls me to him.

"Not a chance, baby. Let's get the girls, get some rings, and go find ourselves an Elvis together. Until you're legally my wife, I'm not letting you out of my sight."

"Ahh . . . I love the sound of that."

He moves in front of me and bends his knees. "Hop on, princess. Let's go get married."

Oh. My. God.

I'm getting married.

"Easton," I gasp as realization dawns. "Oh my God, E. I'm so sorry."

He gathers me in his arms and holds me close. "You still want that annulment, baby?"

A sob bubbles up in my throat. "I was awful to you."

"You didn't remember," he tells me as he buries his face in my hair and runs his hand along my spine. "I knew how you felt. I just needed you to remember it."

"But I . . . I blamed you. I forgot . . . I forgot everything." My chest tightens, and the all too familiar heavy ball of anxiety starts to grow in my chest. "I'm so sorry. I don't have any better words to give you."

"Baby, I don't care about your words. Give me you, and I'm good." He plays with the rings I haven't taken off since he put them on my finger. "Marry me, Madeline Kingston."

I wipe my face and run my hand over his heart. "I just told you I remember marrying you, E."

"Do it again. Do it the way we should have done it the first time. In front of our families. Our friends. You in a white dress. Let Dixon walk you down the aisle and give you to me."

The anxiety shifts and changes course as I think about my family and the way they've vilified Easton. "I'll marry you again whenever you want. But I don't care if anyone else is there, E. I just need us."

"That's not true, princess. You need your family. You

might not be happy with them at the moment, but you can't give up on them. I'd do almost anything to have my mom back." Then almost like an afterthought, he adds, "She would have fucking loved you."

"Fine. I'll try to talk to them again when I get home, but I make no promises." I wrap myself around my husband and lay my head on his chest, almost scared to close my eyes. "E?"

"Yeah, baby?"

"Till we're old and gray."

The Philly Press

KROYDON KRONICLES

COULD IT BE TRUE LOVE?

Guess who was spotted in Washington with her hot hockey hubby? If you guessed baby Kingston, you guessed right. She's still sporting a big, sparkly rock on her left ring finger. And judging by the pictures this reporter saw of the two of them skating on an outdoor rink like they were in some kind of Rockwellian painting, I'd say these two are more in love than ever. I don't know, peeps. My money is still on #babywatch. Should we poll possible due dates?
#KroydonKronicles

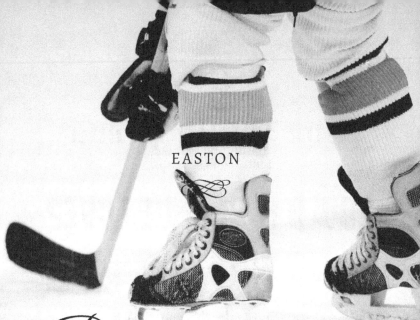

EASTON

"Princess," I whisper as I run my fingers through Lindy's soft hair, a sexy, exhausted smile stretched across her beautiful face. "I've got to get back to my room and grab my stuff. I've got a walk through to get to."

"Hmm . . ." She reaches up and grabs my shirt, trying to pull me down with her. "It's too early to wake up. Come back to bed. I'm not done with you yet."

"Have I created a monster?" I tease. We reached for each other all night last night.

"Baby, you're gonna be sore." I press a kiss to her forehead. "I left you a bottle of water and ibuprofen on the nightstand."

She pouts and sits up with the white sheet covering her gorgeous body. "I won't get you back for two more nights. I'll be fine. Are you sure you don't have time?" Her stormy eyes sparkle bright blue for me.

"You're bad. The bus leaves in fifteen minutes. I've got to go."

She drops the sheet and goes up to her knees, wrapping her arms around my neck.

"Fine. Go to work." She smiles, and I swear my heart constricts like in one of those stupid romantic comedies Jules used to make us watch all the time. "Can I come to your game tonight?"

"You kinda own the team, princess. Pretty sure that's up to you." A lesser man would definitely be intimidated by the power my wife holds. But I'm no lesser man, and I want her to embrace her legacy so she can hold the reins to her own future.

"I do, don't I?" She smiles and presses a kiss to my chin. "I guess that means I have to call Max, doesn't it?"

"Probably not a bad move. It's not like he's been with us for this stretch, but I'm sure he's gonna hear about it if you show up tonight." I run my hand down the smooth skin of her back and cup her perfect fucking ass. "Text me later and let me know how it goes."

She skims her lips along my jaw, and my cock presses against the zipper of my jeans. "Baby, you keep doing that and your brother's going to kick my ass when I get on the bus with a boner."

She pulls back with such a sexy fucking pout on her lips. "Fine. But you know you don't need to let him kick your ass though. You're bigger, and I'm pretty sure you're stronger than Jace."

"You asked me not to hurt him, so I haven't," I admit. "Either he'll get over it or he won't. But for now, at least he can't bitch that I hurt him. He's your brother. How we handle him is up to you."

Her face softens, and she grips my shirt. "Damn, you're one sexy man, Easton Hayes. You've got this whole insanely hot balance of growly dominance and supportive strength going on, and I'm pretty sure I've never seen anything more seductive in my life."

"Don't say shit like that when you're naked in front of me,

baby. I've gotta go to work." I lift her up by her ass and drop her back on the bed, then lean over her. "We leave tonight right from the game, then have one more game tomorrow night before we fly home. You gonna be waiting for me when I get back to Kroydon Hills, wife?"

"Oh, you bet your sweet ass I will, husband."

I wrap her hair around my fist and tug her to me. "I'll see you tonight."

Lindy moans and presses her mouth to mine. "You better win for me tonight, hockey boy."

I force myself to take a step back. "I'll do my best."

I'm sitting on my bench, lacing up my skates before our walk through when Jace comes around the corner and hits my boot with his hockey stick. "Where the fuck did you sleep last night, asshole?"

"Come on, man. Let it go," Boone tells him and tries to move Jace away, but my brother-in-law has gotten himself way too worked up for that. He's looking for a fight, and I'm so fucking over this shit.

I stand to my full height, which is about three inches and thirty pounds bigger than Jace Kingston, and cross my arms over my chest. "What's your fucking problem, Kingston?"

"You're my fucking problem, Hayes. You got Lindy drunk and took advantage of her in Vegas. Now you're married without a prenup—pretty fucking convenient. And now it's your first fucking away stretch, and you didn't sleep in your room last night. So what, you already out fucking some whore?"

I see fucking red and swing at Jace's jaw before I even realize what I'm doing.

Pretty sure I hear a crack too, as he takes a step back to steady himself, but I grab him by the front of his jersey and yank him toward me.

Boone grabs me as Malcom grabs Jace.

"You wanna do this here, man?" I reach out and shove him back, fighting to free myself from the bear-hold Boone's got me in. "Fine. Let's do it. I married your sister in Vegas, and I'd do it again. I don't give a shit about a prenup because I don't care about her money. If she asked me to sign something today, I'd do it in a fucking heartbeat. I've already got more money than I'll ever need."

He shoves Malcom off him and comes at me again. "Where the fuck did you sleep last night, Hayes?"

"In bed with your sister. She flew in yesterday. Now get the fuck out of my face."

Coach Fitz walks into the room, and everyone takes a step back.

"What the hell is going on here?" Fitz booms, and everyone around us stops talking and moves. "You two. With me. Now."

We follow Coach into the room Brynlee and Mason, our other physical therapist, are currently working in. "Can you guys give us the room, please?"

Fitz sounds calmer than I think he actually is.

But that's yet to be seen.

He waits until the door closes, then his face turns purple. "What the fuck is wrong with you two? Jace. You're a captain. Act like it. If you want to keep winning games, we need a goddamned goalie. Preferably one who can stop the puck. Max spun shit into gold and got me one of the best goalies in the whole fucking league. And you haven't stopped fucking with him all goddamned fucking week."

Jace looks at me and then back to Fitz, who I haven't heard curse once in the week I've been with the team.

"Is this hazing? Because no team of mine is going to haze the rookies or the fucking trades. Even if they came from a rival team."

"It's a family thing, Coach," Jace answers him, and I scoff.

"You got something to add, Hayes?" Fitz demands more than asks.

"I'm not your family, Kingston. I love my wife, and she's your sister. You should try talking to her instead of making me your problem." I turn back to Coach and mask the anger I can't quite seem to shake. "I'm here to win games, Coach."

"So help me God, if this shit spills over onto my ice, I'm benching both of you. Do you understand me?"

We both answer, "Yes, Coach."

"Good. Now get out of here and warm up before I decide to bench you for good measure." Then he starts mumbling about stupid fucking asshole players, and I start to wonder if maybe we broke Fitz.

Jace and I get the fuck out of the room and let the door slam shut behind us before he stops in front of me. "I want to tell you to stay the fuck away from me, and we'll be fine on the ice. But you said something back there, and I gotta know."

I brace myself for whatever bullshit he's about to sling my way.

"You love her? Lindy? You love my sister?"

"More than my own fucking life, asshole," I answer him and don't stick around for him to say anything else. I've got a game to prep for, and this isn't how I want to do it.

Lindy

*S*on. Of. A. Bitch.

I'm not sure I can accurately measure how much I don't want to call Max. That's the only reason I can come up with for waiting until I'm in the arena's parking lot, thirty minutes before puck drop to make the call.

"Madeline," he answers after one ring. "Are you okay?" Max always goes straight into protector mode. If any of my older brothers assumed the father role after our father died a few months before I was born, it was Max. He's my oldest brother. My oldest sibling. He's always taken it on his own shoulders to make sure all nine of us were okay. Me especially, considering Jace is my closest sibling in age, and he was in high school when our dad died.

"I'm fine. I'm great, actually," I add, trying to sound cheerful instead of like I'm about to walk the plank. Which is kinda what this phone call feels like. "I decided to fly out for Easton's game tonight and wanted to give you a heads-up that I was here. There were a few reporters at the hotel last night and today, so I'm sure it will end up somewhere."

"Why didn't you call me? I could have arranged a box for you. If you give me a few, I can get my assistant—"

"That's why I didn't call, Maximus. I wanted to come to the game and watch my husband play. Not come and check out the team I own. There's a difference," I tell him, silently willing him to understand.

"That's where you're wrong, kid. There's no difference. It's two sides of the same coin, and you're going to have to learn how to balance it. Married or not, you'll always be a Kingston, Madeline. You're a shareholder in King Corp. An owner of two of the biggest sports teams in the country. You won't ever be just a player's wife. The press will never leave you alone." His tone is short but not mean. Not exactly loving either. But hey, at this point, it's a start.

"And as a Kingston, you need to start acting like one. The press is going to follow you when you show up to a game. You know this. They've been doing it since you were a teenager. And that was before and after the Olympics."

Yup. I'm still, apparently, a disappointment. "Listen, I think I understand what you're saying. And I'll try to handle it better. In my defense, I haven't exactly been married to a player before, and this is all new to me. But that advice goes both ways. If I'm willing to try to see your point, it would be really nice if the family could try to see mine."

"Everyone loves you, kid," he tells me, like that makes it better.

"But that's the problem, Max. I'm a grown woman, not a kid. You all seem to forget I've had the benefit of watching you all go through hell to get your happily-ever-afters. You also forget that I'm very good at math, and I have an excellent memory. Your wife was twenty-three when you met her. And you were older than my husband. So you may want to back off."

He sits quietly on the other end of the phone while I roll down the window of the town car and watch the people all file out of their cars and into the arena. "Listen, Max. I've got to go. The game's going to start soon."

"Point taken."

"What?" I ask him, confused.

"You're right. Daphne was twenty-three when we met, and not everyone was onboard with our relationship."

"And did you give her up?" I ask, already knowing the answer.

"If I had, we wouldn't have Serena."

"Nope. You sure wouldn't, big brother. Think about that while I'm thinking about what you said. I really do have to go though. I want to catch some of the warm-up."

"Love you, Madeline. You sure you don't want me to get you into the box with the staff?"

"No, thank you. I got myself a ticket right by one of the nets. Talk soon." I end the call, and for a brief moment, I think maybe, just maybe, there's some hope with my family.

That's forgotten the minute I get out of the car and a million flashes go off in my face.

"Madeline, look over here."

"Lindy, is that Easton's jersey?"

Lindy, are you pregnant?"

"Did you have to get married?"

Lindy. Lindy. Did you marry him to get him traded?"

Questions fly at me from every direction, and the security at the VIP gate ushers me through. Shit. It's never been like that before. I show the guard my ticket and lanyard. Max thinks he's the only one capable of making a call. But there was no way I was going to my first game, watching my husband play for my team, without pulling a few strings.

I'm escorted through the cavernous halls and brought to the Revolution's bench, where Brynlee stands with Mason, the head physical therapist. Her face lights up when she sees me, and I rush over to hug her.

"I'm so excited you made it." She points next to herself. "You know Mason, right?"

"Yes, we've met." I smile at Mason and offer him my hand.

"Nice to see you, Miss Kingston."

"Actually, it's Hayes now," I correct him with such a giddy smile, I can't hold it back when I look at Bryn.

Her eyes widen, and she turns to Mason. "Do you mind if I . . ." She points to me.

"Go ahead. We're fine. Just don't go far."

She takes my hand and moves down to the other end of the bench behind the sin bin. "Umm . . . okay, Mrs. Hayes. I think you need to spill the deets. Does this mean everything went well last night? I mean, I read the Kronicles this morning, and the pictures they had of you two last night looked hot. Like seriously hot. But when I didn't hear from you all day, I wasn't sure."

I look out onto the ice and immediately find Easton stretching. "Oh yeah. Things went really well. It was amazing, Bryn."

"Amazing enough to introduce yourself as Mrs. Hayes? Does that mean you're done asking for an annulment?"

"I remembered everything, Bryn. *Everything*. So yeah, no more talk of annulments."

She links her arm through mine and looks out over the ice at the team. "So you guys are good?"

"Yeah." I smile, thinking about last night. "I think so. We kinda went about it backward. But I think it's going to work for us."

"Okay. Then you need to deal with the family when you get back to Kroydon Hills."

"Come on, can't you let me just enjoy tonight? I've already dealt with Max. Let that be enough for one day," I beg, not in the mood for another lecture.

"Lindy . . . Easton hit Jace in the locker room. I don't know exactly what happened, but I had to make sure Jace didn't have a cracked jaw. And I totally heard Coach Fitz give Jace and Easton hell. It was bad."

Just then my brother skates over to us and bangs against the boards. "Hey, sis."

"Jace Joseph Kingston. You dick. Why's your jaw bruised?" I demand, and the fucker skates away from me backward with a shrug, like he can't hear me. "I'm gonna kill him."

The music changes, and the guys start to skate over toward the bench.

Brynlee squeezes my hand. "I've got to get back. You need help finding your seat?"

"Nope. I'm good. See you at home when you get back."

"K."

I take a step back as Easton smacks the glass between us. I kiss my hand and line it up with his. "Kick ass, baby."

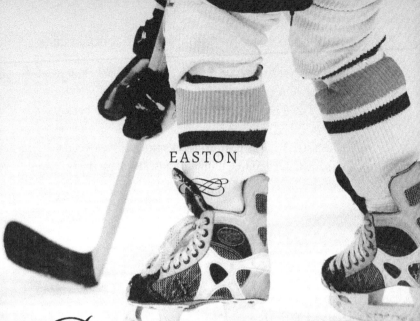

EASTON

"Dude, your wife is a fucking smoke show, Hayes." One of the younger guys on the team taps my stick as we skate out into a line for the national anthem, his eyes locked on Lindy, and I can't even be annoyed. She's fucking gorgeous, standing behind the goal, her eyes locked on me.

"Watch it, asshole." Jace glares when he stops next to me. "That's my little sister."

"Shit. Sorry, Cap."

Jace ignores him and turns my way. "She's wearing your jersey."

"Yeah, she is." I don't even care if it pisses him off.

"She's worn my jersey since the day I was drafted," he grumbles. But something about it doesn't sound as pissed now as it did earlier.

I glance over to him as the singer moves to center ice. "Husband trumps brother."

"Not even sorry about that, are you?" he taunts.

"Not even a little fucking bit," I bite back just before the anthem starts, and we all stop talking. Time to win a game.

There's an electricity in the locker room after our win that's ratcheted up a notch by the fact it was a shutout against a team that's been killing it all season. Reporters are looking to get soundbites they can take back, and I'm not in the mood to talk just yet. I'm new to this team and don't need any egos getting bruised if they try to make the win all about the shutout. But even taking extra time in the showers doesn't do the trick this time. As soon as I walk over to my locker, one of the reporters I recognize stops me.

"Easton. How does it feel to win your second game with the Revolution? Do you think the team is gelling?"

"Yeah, man. We're gelling, and it feels great. I'm just trying to find my place among the incredible players on this team. So far, so good," I tell him and reach for my bag, ready to get dressed and get out.

"There have been rumors that you and the captain, Jace Kingston, have some bad blood between you. Any truth to the rumors?"

Before I can answer him, Jace joins in and throws an arm around my shoulder. An arm I can't shrug the fuck off in front of a camera. The two of us stand there—me with a still noticeable bruise from my black eye and him with a fresh blueish-purple bruise on his jaw from earlier.

"Alex, Alex, Alex," Jace placates the reporter. "I've known Easton since he was in high school. He's a damn good goalie. Pretty sure tonight speaks for itself. There's no bad blood between us."

"So, Jace, tell me. What did you get Easton and your sister for their wedding then?" Alex pushes with a slight edge to his voice. He knows we're full of shit. And he wants to be the reporter to prove it.

Jace laughs and looks at me, suddenly serious.

"I gave them my blessing, Alex. Now get out of here so we can get dressed and on the bus."

Alex turns around to his camera man. "You heard it here first, folks."

As soon as the camera is off, Jace yanks his arm away and shoves my shoulder. "You fucking hurt her, and I will kill you, Hayes. You hear me?"

"Oh, right," I mock him. "Something new. Gee, thanks."

Jace stomps away like a bratty toddler, and I get dressed as fast as possible, then grab my shit to get on the bus. Traveling after a game sucks. My body aches, and the last thing I want to do is sleep on a plane.

I'm expecting the reporters and fans when I walk out of the locker room. What I'm not expecting is to nearly get tackled by a five-foot-two, hundred-pound blur as she throws herself at me.

I drop my bag and grab Lindy as she wraps her legs around my waist and her arms around my shoulders. "You were so good out there tonight, hockey boy."

Her mouth crashes over mine, and our tongues collide. I take two steps forward and lean her against the wall as loud clapping starts thundering in the background. I pull my head back and rest my forehead to hers. "Damn, princess. I might need you to come to every game if you're gonna greet me like that."

She nibbles her bottom lip, then kisses me again, softer and slower. "I have to fly home tonight, but I'll be watching and waiting tomorrow."

"Waiting for what?" I ask, intrigued. I'm not ready to let her go. Not when she feels so right.

"For you to come home," she whispers like it's the most natural thing in the world, and damn, I like the sound of that.

"Come on, Hayes. The bus is leaving," Boone tells me as he walks by.

We don't get so lucky with Jace.

He stops next to us and clears his throat. "Can I talk to you for a sec?"

Lindy looks around, playing dumb. "I'm sorry. Are you talking to me?"

"Madeline . . . please." The words are quiet but strong.

She looks at me, and I lower her legs to the floor. "Go talk to your brother, princess. And call me when you land." I drop a kiss on her head, and I hold her close as long as I can, then glare at Jace. "Don't fucking hurt her."

He gives me a quick nod, then wraps an arm around Lindy to guide her through the crowded hall.

Goddamn. I love that woman.

Lindy

I look over my shoulder to find Easton watching me walk away, and okay, maybe I add an extra little sway to my steps. Then I laugh at myself. Who am I kidding? He's not watching my ass. He's staring at his name and number on the back of my jersey.

Oh, I'm so cashing in on my promise to greet him in this and nothing else tomorrow night.

"Madeline. Watch where you're going," Jace snaps as he opens the door to a small room off the locker room they just exited.

I turn around once the door closes behind us and shove my brother. "Twice my size or not, Jace Kingston, I will kick your ass if you ever lay a hand on my husband again."

"What the hell, Lindy?" He takes a step back and eyes me like I'm a feral cat.

And you know what? Maybe I am.

"You hit him," I whisper-shout, not wanting the rabid press outside those doors to overhear us.

He points to his face like a little tattletale. "He hit me too."

"You hit first." I narrow my eyes at him. "Since when are you a bully, Jace?"

"Since he married my baby sister," Jace huffs out, and I take an angry step toward him and enjoy the look on his face as he backs up. Good. He should be scared of me. Serves him right.

"Do you hear yourself? I'm not a baby, Jace. We got married. Get over it. It was what we both wanted. He didn't take advantage of me. I promise you, he didn't. I swear to God, I don't understand why everyone thinks he did. It was my idea, for fuck's sake. Why is everyone so mad? Why do they all think he's going to hurt me? I trust him, Jace, and I don't trust anyone." I squeeze my fists at my sides, trying to calm the building fury. "He didn't hurt me. I'm not sure Easton Hayes ever could. It's not in his DNA."

I take another step toward my brother, more anger and hurt urging me on. "And if anyone . . . and I mean anyone in our entire family had bothered to ask me, I would have told them all it was what I wanted. What I've always wanted. He's it, Jace. He's always been it. So unless you all want to push me away, you need to get over this shit with Easton. He's on the team. He's signed a contract, and there's a no-trade clause. So he's here to stay. And that's a good thing because where he goes, I go. He's not hurting me, but you are."

I cross my arms over my chest and wait for that to sink in, so utterly over this argument, even if it's the first time I've finally been able to say all this and know, for a fact, I mean every word of it.

"Madeline . . ." Jace's voice softens, and his shoulders drop. "Do you love him?"

"With my whole heart and soul, Jace. He didn't just save my life eight years ago. He's saved it a million times in a million little ways since then."

A muscle ticks in Jace's jaw. "Why the fuck couldn't you have waited to do it with all of us? You could have at least dated for a while. Seriously, let us all get used to you being an adult and shit."

My brother is a lot of things. Eloquent is not one of them.

"Maybe because I did what felt right to me. I married him *for me*. It had nothing to do with any of you." Exasperation mixes with anger and creates a dangerous cocktail in my blood. "You're all so into each other's business, but you've never given me that. *None of you.* You want to control my life. All of you do. But it doesn't go both ways for me the same way it does for the rest of you. I'm not an equal in your eyes. Not for any of you."

When he doesn't say anything, I wait and watch the emotions play out on his face.

He knows I'm right.

"Jace . . . I've been through hell, and I've come out on the other side a pretty well-adjusted woman. But you know what I've spent my life doing?"

He doesn't say anything, so I push harder. "Do ya? No . . . No guesses?"

He shakes his head, and I laugh a soundless laugh. "Funny. Because it revolves around all of you. I did what you wanted. What *all* of you wanted. *Gonna have to train harder to go to the*

Olympics, Lindy. Oh, Lindy. You've got to get a degree. You can balance it. Madeline Kingston. You've got a seat on the King Corp. board. You've got to be at the board meetings. Oh, and don't miss any Kings or Revolution home games. You don't need a life. Family first. Does any of that sound familiar? Because it's the stuff I've heard from all of you my whole life. It's like you all thought you needed to fill in for—"

"Dad," he finishes my sentence for me.

"I never knew him, Jace. He doesn't mean anything to me. He's the man who cheated on my mother and died doing it. I didn't need him."

"He wasn't a bad man, Madeline. He was just bad at love. At least, romantic love. He was great at loving his kids, and even before you were born, he loved you. He was so excited when your mom announced she was pregnant," Jace tells me almost wistfully. "I wish you had a chance to have him in your life. He would have done a better job than we did."

My heart stutters as emotion swells in my throat. "But that's the thing, Jace. I didn't need him because I had all of you. I never felt like I was missing anything. I didn't hate that you all thought you knew better than me. Not when I was a kid. And I'm not stupid enough at twenty-three to think I've got everything figured out. But I'm also competent enough to know what I want and what's worth fighting for. I'm lucky enough to be loved and smart enough to hold on with both hands and fight for it. So I'm going to need you to back off or get out. Those are your options."

Holy shit.

I think that was all a lie. Because this incredibly strong woman I'm projecting right now is shaking like a leaf on the inside, wondering where in the hell I ever got the courage to say all that.

Jace throws his arms around me in a hug that feels like it

might squeeze the life out of me. "When did you grow up, baby sister?"

I close my eyes and fight back the tears I know are right there, burning behind my lids. "Apparently, when you weren't looking, big brother."

He runs his hand over my head and squeezes tighter. "Did I ever tell you *that night,* after the cops left and we were all at Hudson and Maddie's house . . . you and me and Easton were in the family room, while your mom was in the kitchen with the others. The doctor had given you a sedative, and you fell asleep on the couch. Your head was leaning against Easton's chest. And I swear, I don't think he even took a deep breath because he didn't want to wake you up. We were down there for fucking hours, and he wouldn't let anybody move you. When Ashlyn wanted to put you to bed, I stood up to pick you up, but he wouldn't let you go and carried you to one of the bedrooms himself. He was nineteen, Lindy, and I think we all knew then that nothing would ever be the same between you two."

I step back so I can see his face, so hurt and confused. "Then why? Why is everyone reacting like this?"

"It's not fair, but maybe we associate Easton with the night we almost lost you. You've got to understand you weren't the only one traumatized that night. With our family, it's always been us against the world, and we almost lost you. You want us to cut you some slack, but you're going to have to do the same. Maybe we all held on too tight after that, but Lindy . . . you're ours to protect. We circled around you."

A fist pounds against the door, followed by a distinctly pissed-off male voice. "Let's go, Kingston. You're not on that bus in five, it's leaving without you."

"Fuck off, Smitty. I'll be there," Jace calls back, then wipes the tears from my cheeks. "Maybe we held on too tight."

"I've got to be able to breathe, Jace. I deserve to be treated as an equal. I earned that." My heart tightens in my chest, unable to believe I'm having this conversation here. *Now*. In a smelly locker room.

"You have. And I'm sorry. I guess I forgot for a few minutes how much he's always cared for you."

"He loves me, Jace." My voice shakes, but I will not break.

"Yeah, I know. He told me earlier. I guess . . . Well, I guess I just forgot that for a minute."

"Try to remember, big brother. Because this man who you all want to make out to be a bad guy . . . he's already got abandonment issues, whether you realize that or not. You of all people should get that. Both of you had moms die when you were young. Only, unlike you, he never had a dad to lean on. He had Kenzie and eventually Jules and Becks. And he thought he had all of you, but he was shown just how wrong that was as soon as he got traded, came home, and all of you decided he wasn't worthy of your family."

"Madeline." Jace looks horrified.

Good. He should be.

"That's not . . ." He trails off, and there's another bang on the door.

"Time's up, Kingston."

I wrap an arm around my brother. "I've still got to battle it out with the rest of them, but it would be really nice to have an ally. And maybe while you're at it, give my husband the apology you owe him."

"I'll see what I can do," he tells me as he grabs the doorknob.

"Jace," I stop him. "Do better."

THE KEEPER

KENZIE

I'm kinda glad dinosaurs are extinct because I'm pretty sure I'd try to ride one after a few too many cocktails.

BRYNLEE

Uhhh. Did Everly take Kenzie's phone?

KENZIE

Nope. It's me. I'm just drunk.

EVERLY

I don't wanna ride dinosaurs when I'm drunk, ladies. I have much better things to ride. You all should try it.

GRACIE

Pretty sure Lindy's the only one who hasn't tried IT, yet.

LINDY

Well . . .

EVERLY

Holy hell, Hayes. Did you climb that tree?

KENZIE

I didn't climb any trees.

EVERLY

Not you Hayes. Lindy Hayes. Like Mrs. Hayes. The one who just told us she banged your brother.

KENZIE

When did she say that?

BRYNLEE

How drunk are you?

GRACIE

Better question – where are you and who are you with?

EVERLY

Uhh . . Best question. How was it?

KENZIE

It was too many dirty martinis. They were very, very dirty.

GRACIE

Kenzie - where are you? I'll come get you.

LINDY

Just checked the app. She's at West End.

KENZIE

I'm at Maddox's bar with some friends from school. I'm fine.

BRYNLEE

Dude. She has other friends?

EVERLY

Do we do other friends?

GRACIE

You don't because nobody else wants to be your friend. Sorry not sorry, sissy.

EVERLY

I'm gonna smother you with my pompom, sissy.

KENZIE

Somebody woke up and chose violence today.

LINDY

My plane's about to board. I'll be home either really late or really early, depending on how you look at it. See you tomorrow.

EVERLY

Drinks at West End tomorrow night so you can fill us all in on how last night went?

BRYNLEE

No fair. I won't be home until after midnight tomorrow.

GRACIE

You saw her today.

LINDY

Sounds good. See you tomorrow.

The Philly Press

KROYDON KRONICLES

HOCKEY HUBBY STAKES HIS CLAIM

I'm absolutely giddy from the pictures I have to share with you today, peeps. Baby Kingston in her hot, hockey hubby's jersey, wrapped around said hottie like a koala bear. Check out the second pic, people. There is definitely ass grabbage happening, and I'm here for it all. Did anyone else notice how baggy that jersey is, folks? Could it be hiding something like a bump? Forget #babywatch. I think we need a new hashtag. #bumpwatch. Go forth and let me know if you spot one before I do.

#KroydonKronicles

LINDY

Red-eyes suck. Especially when you're losing time, and Seattle is three hours behind Kroydon Hills. So there's three hours of my life I'm never getting back. Add to that the extra hour we sat waiting on the tarmac before we took off last night, and I'm cranky, exhausted, and not at all in the mood to deal with the reporters following me through the airport.

But holy shit.

They've got nothing on the ones flashing cameras in my face when I step outside.

It's a madhouse as I try to make my way to the massive SUV Uber waiting at the curb.

I ignore them as best I can. This isn't the first time everyone has wanted a picture or a comment, and it won't be the last. But it may be the first time they've been this intrusive. I'm used to Charles being here to handle it. Guess that's what I get for exerting my independence. I might be regretting that one right about now.

I nearly trip as a camera is shoved in front of my face, and

I stumble to open the back door. With shaky hands, I steady myself, slide in, and slam the door shut.

The driver turns around. The smell of weed mixes with a nasty air freshener, like that's going to mask it. "You a celebrity or something?"

"I'm a figure skater," I tell him and buckle my seat belt. This guy doesn't need to know I'm a Kingston, and I'm not about to advertise it.

Shit. If that thought doesn't make me realize maybe I do need some form of security, I'm not sure what will.

The driver confirms my address, and I shoot off a text to my sister Amelia's husband, Sam, asking if he's got time to talk today. I'm willing to at least discuss security if it's on my own terms. If they work for me, I can tell them to back off when I need space. They'll answer to me, not my family.

The city streets are empty as we make the quick drive from the city back to Kroydon Hills. It gives me a chance to get my bearings before the driver, thankfully, pulls into our building's underground garage to let me out. As I open the door, he scoffs, "Didn't know figure skaters got paid enough to live here."

Eww.

I refuse to dignify that shitty comment and shut the door. "Thanks."

He's definitely not getting a good review.

I smile at our doorman and consider stopping in the coffee shop but decide sleep trumps caffeine this morning. Elevator it is. My bed is calling me.

But when the doors open, I'm on the sixth floor, not the seventh, and Kenzie is waiting to step on. She looks at me and closes her eyes. She's a hot mess. Messy hair. Smoky eyes smudged, but her day old-mascara still looks half decent. What the hell?

"What are you doing down here?" I ask as I hit the button to close the doors.

Kenzie's head thunks against the wall, and she shushes me. "Not so loud," she whispers.

Ok-ay. Guess it's her turn for the hangover from hell.

This day is off to a stellar start.

The two of us ride up to our floor in silence, then pass Gracie in the hall. "Hey. You're back," she smiles, then looks Kenzie over like she smelled a skunk.

"Yeah. Just got in. You off to class?" I ask.

"Yup. Baby ballerinas at Mom's studio. We still on for drinks tonight?" Grace asks, and Kenzie groans and shoulders past us into the condo. "What's her problem?"

"No clue. Did you see her last night?"

Gracie shakes her head. "But judging by the look of her, I'd say she either had a really *good* night or a really *bad* one. It could go either way."

"Yeah. Guess so. I'm gonna go crash. I'll see you tonight."

She moves to the elevator, and if I had the energy to run, I'd sprint to my bed. Myrtle greets me when I walk through the door, and I give her some loving and a treat, then let her follow me to my room. She uses her doggy stairs to get on my bed, and then my lazy dog passes out before I do.

The next time I wake up, it's because Everly is sitting on my bed, laughing.

I crack an eye open, close it again, and rub both eyes with my fists.

"What are you doing here?" I grumble and push my hair out of my face, then wipe the drool from my mouth. I'm not what you'd call a *pretty* sleeper.

Evie laughs at something she's reading on her phone and leans back against my pillows, smiling. "The game starts in an hour, and I thought you'd want to shower before we hit up West End." She looks back down at her phone and laughs harder. "You should see this shit. The Kronicle is doing a bump watch and polling for an *It couple* nickname. My fave is Hazy. Get it? Hayes and Lindy—Hazy."

I grab my glasses from my nightstand and force myself to sit up so I can see what she's looking at. "Wait . . . did you just say *bump watch*? Like they think I'm pregnant? Did they get another fat picture? Jesus. One fucking burrito and everyone thinks I'm pregnant."

"They've been speculating the quickie wedding was because you're pregnant. Have a few drinks at West End tonight. That should put it to rest. Some asshole will snap a pic and send it in."

I snatch her phone and look at the screen. There's over two thousand comments on the last post. "Two thousand people are discussing whether I'm pregnant because my jersey was big last night?" I shake my head and toss the phone back to her, then lie back down. "It's a jersey. They're big."

Everly gets up and yanks my blanket off. "Get up. Get showered. And let's go."

I look at her and wish I had something to throw. "You should have been a drill sergeant."

"Camo is not my color. Now move your ass."

"Should you be drinking, trouble? Rumor has it you're pregnant with twins."

I take my tequila and club soda from Maddox with a glare. "You're a dick."

"He can't help himself. It just comes naturally." Everly sips her white cranberry cosmo and leans back against the bar, her eyes trained on Gracie, who's flirting with a guy we've never seen here before. "Who is that?"

"No clue, but he sure is pretty."

"Dudes don't like to be called pretty, trouble." Maddox holds up the remote and changes the channel. The Revolution comes on the screen, and they zoom in on Jace and Easton talking down by E's net. "They making nice yet?"

"Not as of yesterday. But I think Jace and I came to a tiny understanding." I pinch my fingers together a smidge, and Maddox ignores me. "I guess we'll see."

I shoot off a text to E, telling him to kick some ass tonight, then scroll through my messages. "Shit." I slept through Sam's text.

"Something wrong?" Maddox prods, like the nosey little shit he is.

"I messaged your dad earlier and missed his text when he got back to me."

"Looks like you missed him till Monday then. He's taking Mom away for the weekend." He looks around the bar, then back over at me. "Where's Kenzie? She feeling okay?"

"Huh?" I ask, then realize what he said and focus on him. "Kenzie had a study session tonight with some friends from school. Were you here last night? She looked like shit when she got in this morning."

"Yeah. She was hammering shots. She got pretty wasted."

Gracie moves across the room with her mysterious stranger in tow. "Can I get another lemon drop, madman?"

"Sure. Anything for you, buddy?"

The stranger shakes his head, and I cringe. If Maddox calls you *buddy*, you're a douche, and that's his way of telling us to stay away.

That's about right. Gracie would find the douche tonight.

They're always attracted to the quiet ones.

By the time the first period of the game is over, I'm ready

to scream. Every time these asshole commentators get a chance, they're bringing up Easton and me.

How they think he feels playing for a team his wife owns.

How there were rumors of a rift between Jace and him.

How well the captain and his goalie seem to be working together tonight.

"Jesus Christ. Shut the fuck up and talk about the game," I yell at the TV, and the small group of people here cheer.

West End is a local spot. Sam has always kept it that way, and now Maddox does the same. So when a flash goes off from the other side of the room, Maddox flies across the bar. "You got any clue who you're fucking with, buddy?"

The dumb fuck blinks at Maddox, clearly not having any idea who he's fucking with, then yells when madman smashes his camera and throws him through the door. "Bill me, asshole."

When Maddox turns around, he throws his arms out. "Anybody else in here have any ideas about taking pictures of my family, you might wanna get the fuck out too."

Everyone cheers, and the Revolution scores the first goal of the night. So I figure *what the hell* and yell, "Drinks on me for each goal they score tonight, guys."

Maddox shakes his head as he gets back behind the bar. "Always causing fucking trouble."

I smile sweetly. "But you love me."

"Whatever. Give me your credit card."

I slap it down onto the old cherrywood bar and look back up at the TV to watch my husband. Oh yeah. I can't wait for him to get home tonight.

THE KEEPER

Once we're home after the game, I change out of my jeans and throw on a pair of sweats, then toss a couple of things in a bag and head back down the stairs. Kenzie's in the kitchen, eating leftovers. Her glasses are pushed up on top of her head and notebooks are spread out in front of her. "Hey, we missed you tonight."

She looks up, an egg roll halfway to her mouth. "Sorry. I was at the library late. Where are you going now?"

Big girl panties, Lindy.

Put 'em on and pull 'em up.

"I thought I'd wait for Easton downstairs. Madman mentioned earlier that a ton of E's boxes and furniture were delivered yesterday and today. I thought I'd help him unpack."

"Oh." Her eyes pop wide. "I was going to do that. Does that mean you and Easton are . . ."

I drop my eyes, feeling nervous, but I can no longer hide my smile from one of my best friends. "We're good, Kenz. We're really good."

"Like no more annulment good?" She drops her egg roll and stares at me, waiting.

"No more annulment. He asked me to marry him again. He wants to do something big this time," I whisper. I hadn't told anyone that. Well, no one besides Jace.

"Do you want that?" she asks with hope dancing in her eyes.

"Do I want to be married to him? Oh yeah," I say softly, scared to get too excited yet. "But I don't know about the whole big wedding thing. I'm not even talking to my mom."

"You'll fix it, Linds. She's your mom. You have to fix it. Take it from someone who'll never get the chance to talk to her mom again. *Fix it.* Fight it out, but fix it. Get things straightened out between you and Easton, then work on the stuff with the family, starting with your mom."

I lean my head against hers. "How'd you get so smart, girl genius?"

"Born that way, I guess. We all have our gifts."

"Well, just so you know, things are straight with Easton and me." I grab a vitamin water from the fridge and drop it in my bag, then steal of bite of Kenzie's egg roll. "Yum. That's good."

"Oh my God. Did you remember?" She claps her hand like a giddy cheerleader, and I just smile until she pulls out her phone, and then a text pops up on mine.

"What did you do, Kenz?"

"Watch that with E when you get a chance, okay?"

"What is it?" I ask with a funny feeling.

"Just trust me and watch it with my brother. Now go. I've got another chapter to get through before I can go to bed, and I *really* need to go to bed."

"Okay. See you tomorrow." I slip Myrtle's leash on her collar and take her with me down to Easton's condo.

It's smaller than ours. A traditional two-bedroom with an open layout and a big balcony overlooking one of the small rivers that feed into Kroydon Lake. Movers may have brought his things in, but they sure didn't unpack. There's a ton of boxes in his family room and a new leather couch with the tags still attached up against the white wall.

I decide to explore further and walk into the smaller bedroom. The furniture in here was clearly brought in from his place in Vegas. Worn and loved. There's a wooden desk Easton's had in his room for as long as I've known him. *Office* is written on the outside of the boxes in a handwriting that's clearly not Easton's. I assume he hired a service to pack his place up and move it across the country.

I peek inside the box and pull out a framed picture of

Easton standing between his mom and Jules. Kenzie sits on E's waist, and everyone is cheesing for the camera. It's a great shot, so I stand it up on his desk and smile.

I wander further down the hall to the master bedroom and find a new box spring and mattress sitting on top of a big platform bed. I run my hand over the California king mattress and tear off the tag, then open up the big box next to it and find the sheets.

Looks like it's time to do some laundry.

Two loads of laundry later, I've got his bedroom unpacked. His clothes are hung up, his bed is made, and the little bit of toiletries I could find are on the counter in the master bath. His big bed is covered in flannel, like some sort of lumberjack's. But it's soft and smells like fabric softener, tempting me to lie on it for just a few minutes. I curl up on my side and close my eyes. Just for a minute. Just a little rest.

That's what I told myself. But the already dark room is pitch-black when I feel the bed move next to me before I'm wrapped in my husband's massive arms. His smell envelops me, and I hum, feeling happy. "What time is it?"

"Late," he whispers against my ear, sending warm tingles down my body. "My flight just got in, and I got your message to come here and not your place."

I roll over and rest my cheek against Easton's bare chest. He smells like soap and sandalwood, and my body roars to life as I wrap myself around him. "I missed you."

"You look so fucking sexy in my jersey, baby." His warm lips press against mine, and I sigh and tangle my legs with his. Easton's hand slides up my bare thigh and stops on my bare hip. His eyes heat and darken. "Do you have anything on under my jersey, wife?"

Butterflies take flight in my stomach. I run my fingers down Easton's delicious chest. Heat blooms inside me, emboldening me. "How about you find out for yourself, husband."

He shifts between my legs, and goosebumps dance down my skin, tiny little fires sparking to life everywhere they touch.

He slides his hands up my ribcage and cups both breasts in his big, rough hands, gently squeezing before his thumbs brush over my pebbled nipples.

A bolt of lust tugs deep inside me.

"Easton," I pant, not sure what I want except that I want it all.

"Seeing my name and number on you is so fucking hot, princess, but I need this gone." He shoves my jersey up and over my head. Strong lips press against my throat. Hot and heavy. Licking and sucking his way down to my collarbone. He nips at my breasts, and his teeth scrape my nipples. He's everywhere as he worships my body.

Our heavy breathing becomes the only thing daring to break the silence of the night.

The snow outside practically glows iridescent white from the balcony, leaving us otherwise bathed in darkness.

He shifts down, and I spread my legs when his mouth finally finds its way to my pussy, and he blows a hot breath against me.

I look down at his dark eyes and crooked grin and can't look away.

Easton Hayes is intoxicating.

He gives me a confidence I've never had before, and I love it.

"I want to taste your pretty pussy, princess." And God, that dirty mouth.

I tug his hair and spread my legs, dropping my knees open on the bed. "What are you waiting for?"

He runs his fingers along my sex, and those tiny sparks from earlier grow to full-blown flames as he spreads my lips. Gathering my wetness, then sucking it from his finger.

A chill skips along my overly heated skin, and my back arches off the bed when he dips his finger inside me. "Ahhh . . ." I moan until he steals my breath and sucks my clit into his mouth.

I call out breathlessly, my muscles tightening, my abs quivering.

My thighs clench and my knees lock around his head.

Desperate to be closer.

I shift my hips, needing more and moaning when he gives it to me.

Easton groans and sucks me into his mouth, flicking and kissing. His tongue spears inside me before it goes back to my clit. Pulling me closer, he fucking devours me.

The gentleness from our first night together is gone, replaced by a desperate frenzy that's fanning the flames higher and higher.

With every swipe of his tongue and stroke of those blunt, rough fingers—pushing inside me, stretching me, fucking me —my body heats and shakes. I claw at him. At the sheets. At anything I can reach as my muscles pull taut and my orgasm sits at the very edge of my vision, teasing me.

His rough hands slide under my hips and grip my ass, changing our angle.

Pulling me flush against his mouth, he growls against my sex, and the intensity of the vibrations sends me spinning. "Ohmygod, Easton."

My hips lift, and my body throbs like one big heartbeat threatening to tear me apart.

The pressure builds higher and higher until it's too much.

Too much and not enough at the same time.

My nails score his skin as I moan and gasp and beg him to let me come.

My skin burns as a sizzling pleasure builds to a fucking inferno.

Until it's too much and I think I might lose my mind.

I look down at those dark-green, hooded eyes, locked on mine, and cry out again and again.

Easton drags one finger inside my pussy as another presses against the puckered skin of my ass.

And just when I think there's no possible way I can take any more, his teeth scrape over my pulsing clit. I detonate in a violent orgasm that shakes me to my core.

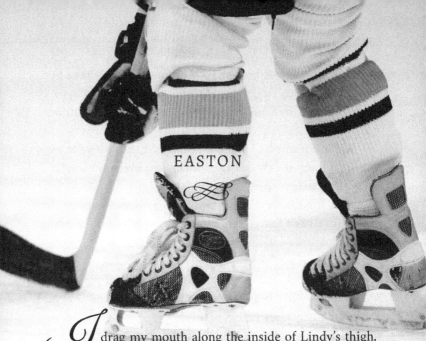

EASTON

I drag my mouth along the inside of Lindy's thigh, nowhere near done with her yet.

I'll never be done. Never.

My fucking intoxicating wife lies motionless on the bed with a sated, sexy, sleepy smile.

Like a woman who's just been fucked hard. But I've barely started.

I drag my lips over every inch of her skin and enjoy the soft sounds of Lindy's whimpers while she basks in the happy afterglow of her orgasm. Each stroke of my tongue lingers and pulls along the defined muscles of her body. Every hard-earned dip and curve remind me what a fighter my woman is. She's an elite athlete. So much stronger than she gets credit for.

A soft hum slips past her lips as I graze my teeth over her hips. I've noticed she likes when I use my teeth. I press a kiss to her delicate abs and swirl my tongue around her flawless tits—a perfect handful—before covering her body with mine.

"Easton," she whimpers. "I think you broke me."

With the taste of her pussy fresh on my lips, I take her mouth with mine. Dragging my teeth over her bottom lip.

She moans into my mouth, and my cock weeps in my boxers.

"You're overdressed, hockey boy." Her hands move to my waistband, and she shoves my shorts down over my hips. "It's my turn to taste you, E."

"Next time, baby. I've been dreaming about being inside you again since I left you in my bed yesterday, and I'm done waiting." I strip off my boxers and take her mouth again. "You ready for me, baby?"

I fist my cock and drag it through her soaked pussy. *Her heat.* Fuck me, her heat is intoxicating. My head spins as all the blood in my body rushes to my dick.

Lindy brings her knees up to cradle my hips between her thighs and then tilts her pelvis. "That doesn't sound fair to me," she pouts.

I run my thumb along her bottom lip as my cock lies heavy between our legs, the tip teasing her hot cunt. "I fucking love you, princess," I growl and lace my fingers with hers, holding them against the bed on either side of her face.

"Then fucking show me. Fuck me, Easton."

Challenge accepted. Keeping her hands trapped beside her head, I tease her clit with my dick and drop my mouth down to one pale-pink nipple. Sucking the tight little peak between my lips, I groan as it grows even tighter.

Her breath hitches in her throat, and I slide through her pussy, teasing her clit. "You with me, wife?"

She whimpers, and I push in.

Just the tip, barely moving.

I pull back out. Teasing her.

Building her up slowly.

"Easton," she cries out as her tight little pussy clenches around my cock in a fucking vice-like grip. "Oh God, E . . ."

I bring my face close to her and graze my teeth over her thrumming pulse. "I'm taking my time, princess. I want to fucking worship you."

"Easton..."

The sound of my name on her pretty, swollen lips is like a drug. She's my drug.

"Fuck, you feel so good, E." She wraps her leg around my waist and drags her wet pussy over me. "So, so good." Her nails dig into my shoulders as her hips slowly circle my cock.

"That's my good girl. Take what you want."

She buries her face in my shoulder and bites down on my neck. A loud moan echoes in the quiet room. My girl likes a little pleasure mixed with pain.

"I want you to fuck me, Easton. *Hard*. I want to feel you for fucking days. Every step I take, I want to feel you." She circles her hips and clings to me.

I shift my hips and pick up speed, bottoming out in her.

My abs contracting with each thrust.

Her hips move in sync with mine, like they were always supposed to.

I drop my hand between us and press my thumb against her clit, wanting her to come for me. Needing it. Feral for her. My wife.

"I'm so close, Easton." She moves faster and gasps as she locks her legs behind my back and shakes. "Oh God. *Oh God.* Yes."

"That's it, baby. Come on my cock. Fucking give it to me."

Lindy screams and shatters around me.

Clinging to me until I lean back on my knees.

"What?" She looks up at me wide-eyed and dazed.

"You didn't think I was done with you yet, did you, princess?" Her face lights up, and man, the sight in front of me could bring me to my knees.

Hunger burns in the depths of her stormy-blue eyes, and

my girl drops those gorgeous legs to the bed so I can pull her up on my lap.

Lindy adjusts herself. Her nipples rubbing against my hard chest. Her nails trailing over my shoulders.

And when she grinds down on my cock, she unleashes a fire that incinerates any thread of soft and slow I had left in me.

"Do you have any idea how long I've dreamed of fucking you, wife? Your mouth. Your cunt. Your perfect fucking ass."

I dig my fingers into the globes of her ass and set a hungry rhythm, fucking away the years of emptiness between us.

The years we wasted.

"Fuck, Lindy. You're so tight, baby."

"So good, Easton. God, don't stop."

My muscles burn as we fuck our way into oblivion.

Her pussy pulses around me, and I lift her off me and flip her over to her knees, then drag my hand over her gorgeous round ass and spank it with a quick snap of my wrist. "Up on all fours, baby."

Lindy pants as she hurries to put her ass in the air. Then she tosses her hair over her shoulder and peers up at me.

Fuck me. My mouth goes dry.

My red handprint glows back at me against her creamy skin. "So sexy, baby."

Her forehead drops to the bed, and I kiss the handprint and drag my tongue over her pussy until my face is fucking covered in her wetness.

She backs up, pressing her ass against me. "Please, Easton. Please."

"Please what, princess?" I pump my cock and drag it through her swollen pussy, loving her desperation and the way it matches my own.

"Ohmygod . . ." she cries out. "Please fuck me, husband.

Please, please, please. I want to come again. Make me come. Please let me come."

I drag my thumb through her wet sex, then circle it around the tight ring of her ass, running it around the puckered skin. Pushing my thumb into her ass, I plunge my cock in her pussy and bottom out.

Lindy shakes and screams a guttural scream, and I wrap her long hair around my fist and tug her up.

Her back arches.

Her ass slaps against my thighs as I pound into her.

"Fuck, baby. You take my cock so well."

I lean over her and kiss my way up her spine, then pull her head back until my mouth is next to hers.

Her body shakes, as she pants and moans and circles her arm back around my neck.

Dragging me closer.

My heart beats a wild, staccato rhythm, pounding against the inside of my chest. Threatening to explode. "Fuck, Lindy. Never again. Never going without you again. It's you and me."

"It's us," she keens and quivers. "It's always been us."

"It's always gonna be us, princess." I take her mouth in a soul-stealing kiss.

She's it. She's everything.

She tightens around me and slams her pussy against me one final time before her orgasm washes over us. Pulsing and pounding. Her body trembling. Vibrating around me as I fuck her through it.

White-hot blistering heat builds at the base of my spine.

Pleasure pulls tight before it rushes through me, and I come with her name a fucking benediction on my lips. I'll only ever worship at her altar.

Aftershocks spark through us as Lindy lies on the bed, incoherent.

Can't say that doesn't make me feel like a fucking man.

I climb out of bed and walk into the bathroom to get her a warm washcloth.

When I walk back in, she tilts her head to me.

"What are you doing, hockey boy? I don't think I can go another round." Her blue eyes smile at me. "Not yet, at least. Give me a few minutes," she giggles. And that is so my girl. I just fucked her through three orgasms, and she's giggling.

I run my hand over her thigh and open her legs. "I know, princess. Just let me take care of you." I gently run the washcloth along her sex before I toss it to the floor, then climb on the bed next to her, dragging the blanket up over us.

She rolls over and drapes herself over my chest. "Promise me it's always going to be like this, Easton."

"Till we're old and gray, baby. I swear on my life." I press my lips to her head.

"Those are the only vows I need, husband."

"I love you, wife."

𝔗𝔥𝔢 𝔓𝔥𝔦𝔩𝔩𝔶 𝔓𝔯𝔢𝔰𝔰

KROYDON KRONICLES

HAZEY FOR THE WIN

Hazey watch is in full effect, peeps. This reporter hasn't seen our favorite newlyweds together in over a week. Could it be? Has the shine already worn off their union? Or maybe are they just avoiding the spotlight and enjoying the honeymoon period. It's hard to believe baby Kingston would have willingly skipped every game her hot hockey-god of a husband has played all week. But she hasn't been to any, home or away. Who knows, maybe she's got royal princess-level morning sickness happening and can't get out of bed. I'm not giving up on them yet. #Hazey #bumpwatch #KroyonKroinicles

LINDY

BRYNLEE

I want a man who's going to wake me up with a mimosa and call me a queen.

EVERLY

Bitch - I can't even get a guy to wake me up with a cup of coffee.

LINDY

Easton woke me up with his head between my legs.

KENZIE

OMG. I just threw up in my mouth.

GRACIE

Swallow, sweetie. I'm betting that's what your brother does.

KENZIE

I hate you.

EVERLY

Are you and lover boy going to come up for air anytime soon? We miss you.

THE KEEPER

LINDY

> Dramatic much? I've been around.

GRACIE

You've slept at his place for a week.

KENZIE

Still hating you, but need to point out, they're married. Pretty sure it's their place.

EVERLY

So what? We've been demoted from roommates to tenants now?

GRACIE

Lindy's right - Dramatic much?

EVERLY

Somebody's gotta be.

BRYNLEE

The family's been asking about you at the Revolution games this week. Your mom came to both home ones.

LINDY

> Yeah. But she hasn't called me.

KENZIE

Easton said Jace has been better.

LINDY

> I think so. He's accepted it.
>
> Easton and I were thinking about getting a Christmas tree today and decorating it tonight. Anybody want to come over and help?

EVERLY

I'm in.

KENZIE

Me too. I'm on winter break for an entire glorious month. Halle-freaking-lujah

GRACIE

I'll be there after Nutcracker practice. Tonight's our first full dress rehearsal.

BRYNLEE

I'm good. No game and no practice mean I've only got a few guys on my schedule today. You need us to bring anything?

EVERLY

Like air freshener to mask the scent of sex in the air? Or disinfectant for all the surfaces you should probably clean before you put any food out?

KENZIE

Why are we friends?

GRACIE

Consider it our good deed. She's mentally unstable. That makes our friendship kind of like charity.

EVERLY

Whatever. I'm just saying what you all were thinking.

LINDY

OMG. Bye. I'll see you tonight.

"*B*aby. The ceilings in the condo are only ten feet tall. This tree is taller than that." Easton pulls my white cashmere hat down over my ears and kisses me. "We need to look over there, where the smaller ones are."

"But I got enough lights and decorations for a big tree." I fake pout and bite down on my bottom lip until my husband groans. Yup. That always gets him going. His thumb replaces my teeth, and he gets all growly.

"Unless you want me to throw you over my shoulder and drag you home right now, you need to stop doing that."

I circle my arms around his neck and press up on my toes. "You're not exactly deterring me, Easton."

"Aunt Lindy, Aunt Lindy," my nephews Atlas and Asher call out to me as they run my way. The eight-year-old twins are followed more slowly by their older sister, Saylor, and teenage brother, Cohen, with my brother Jace and his wife, India, behind them.

I drop my arms and turn to hug the boys.

This is the longest I've gone without seeing my family. But between the drama still brewing and the intense media scrutiny we've been under, laying low has been easier than dealing with everything.

Saylor pushes the boys out of her way and wraps an arm around my waist. The little blonde pixie is the spitting image of her mother, down to her quiet demeanor. "Missed you," she softly tells me, never wanting to bring any attention to herself. Which works for her because her brothers demand every second of it.

Cohen, who at fifteen is already as tall as his father, wraps an arm around my shoulder and squeezes. "You getting this tree, Aunt Lindy?"

"I don't know . . ." I drag out as I look at my brother and his wife. "Easton says it's too big."

India shakes her head and kisses my cheek. "They always think it's too big, but they always make it fit."

Jace coughs, Easton chokes, and Cohen groans and walks away, declaring he's getting hot cocoa.

"Hey, why don't you guys go with your brother? We'll meet you at the car in a few minutes," India tells the kids, then watches them walk away before she smacks Jace. "You're terrible. That is not what I meant. And now my son thinks *it's* too big."

Jace grabs India's ass. "Come on, pretty girl. You know it is."

He doubles over when she throws an elbow at his kidneys.

"I'll stop," he groans out.

I enjoy the way Easton's hand lands protectively on my hip as he pulls me into him. "Enjoying the day off, man?" E asks Jace.

"Yeah. Catching up on some family time. How about you guys?" He looks between us, and I'm pretty sure he's not plotting Easton's death, so that's something.

Easton squeezes my hip, and I lean against him. Well, we've been to three different stores, getting holiday stuff, and now your sister wants a tree that won't fit in the condo. So I'm not really sure how the rest of the day is gonna go."

There's no heat behind his words, and it might make me sound silly, but I love that he's teasing me like this. Like this is just a normal day for us.

"Ha. Good luck with that," India snorts. "Kingstons always want the biggest tree they can get their hands on. It never fits, then half the brothers come over to be manly men and stand around the thing, drink a few beers, and bust out their chainsaws to make it fit in the house. It's tradition."

My heart aches because I haven't seen any of my brothers, except Jace, in weeks.

Easton must sense my sudden sadness because the hand resting discreetly on my hip curls around my waist, and he kisses the top of my head. "Sounds like we better add a chainsaw to our list, princess. If you want this tree, I'll call Pace, and we'll make it fit for you."

"Thanks," I whisper and turn to my brother. "See you guys later."

"Lindy," Jace calls out. "Stop."

When I raise my eyes to his, anger mixes with sadness and threatens to spill over. "You didn't do anything wrong, Jace. It's okay."

"It's not okay. Somebody has to make the first move, and I know you're not going to want to hear this, but you're younger, Linds. It's got to be you. Come to the game tomorrow night. At least then, you'll have to see some of them. It could be a start."

"Maybe," I offer without making any promises.

"I'll be there with the kids," India tells me as she reaches out and squeezes my hand. "Ask my somewhat misguided husband. I've been so excited for you and Easton since I found out you two got married. I always thought you would end up together."

You'd think by now I'd be done tearing up, but apparently, I'm not. "You're the first person in the family to say that."

"Thanks, India," Easton tells her. "We have a meeting with Sam tomorrow to discuss getting Lindy security on her terms. I think we need to make sure we've got something in place before she goes to one of my games. These fucking paparazzi are like roaches. They're fucking everywhere."

"*Really?*" Jace's shocked voice almost makes me laugh—almost. "You're going to give in and get Charles back?"

"Maybe. We're going to see what Sam says. I'm telling you, Jace. I'm not doing this again if anyone other than me is in charge of it. If I give in and get security, they have to answer to me."

"Give in to who?" Jace asks, confusion lacing his tone.

"To me," Easton pulls me closer. "They followed us to the grocery store last week. She canceled her baby skaters classes because she didn't want to leave the damn house. I need her safe." He turns me to face him, and I hate the guilt I see on his face. "You'll have all the control, baby. But you need someone there when I can't be."

"Can we please just get the tree? I was having a good day, and I don't want to ruin it. We'll deal with it tomorrow."

I may know I need security, but I've only just started to get a tiny glimpse of having a life without being followed by a bodyguard, and it felt really nice to not be followed twenty-four seven. Knowing I need one is pissing me off.

"Yeah, princess. Let's go home."

I nod and lean into him. "Let's go home."

"Okay, that's fucking hot." Everly taps her martini glass to mine and sips as we watch Easton, Pace, Maddox, and Callen working to get our tree up. India was right, Easton and Pace had to cut the bottom off to get the thing to fit in our condo. But it was worth it because it's perfect, and now I have an incredible view of my husband's ass as he sets up the tree.

"You realize Callen's over there too, right?" Grace asks as Everly drools over Pace and his arm porn. I mean I can get on board with it. He came from the office in a rolled-up white dress shirt before he started helping Easton. Brightly colored tattoos cover his forearms, which flex and move as he helps Easton adjust the tree.

Kenzie comes in with the shaker of candy-cane martinis

and looks at the guys, then back over to us. "Have you seen him since we've been back from Vegas, Evie?"

"Nope. I came, he conquered. No repeat needed."

I choke on my martini and somehow avoid snorting it out of my nose as Brynlee sighs.

"Seriously, this is better than porn if you can get past Maddox over there." Bryn kicks her feet up on the coffee table. "I need to get laid."

"I volunteer as tribute," Callen announces, then winks at Brynlee, and we all laugh.

That's how our night goes.

Lots of laughter.

Lots of drinks.

Maddox gets someone to deliver from Sam's restaurant, Nonna's, and we decorate my first big Christmas tree in my first place with my husband. Easton corners me in the kitchen while Callen and Evie argue whether the tree needs ribbon or popcorn strands. He lifts me onto the counter, and I wrap my legs around his waist.

"Did you notice nobody knocked before they came in, princess? Not even Callen or Maddox." His nose runs up my neck, and my head drops back against the cabinet behind me.

"Better get used to it, hockey boy. They don't knock. If you want privacy, you better lock the door."

"Don't bother," Maddox tells us as he walks into the kitchen and grabs two beers out of the fridge. He opens them both, then hands one to Easton. "I've got keys to the whole building."

"Wait." I push Easton away and hop down. "I own the building, and I don't have keys to everything. Why do you . . . how do you?"

"I have my ways, trouble. Hear you're meeting with Dad tomorrow."

"You know everything too, madman?"

"Wouldn't you like to find out." He turns and walks away, and Easton looks at me funny.

"Glad he's on our side," he tells me.

"You have no idea." I lace my fingers with his and tug him behind me. "Come on, everybody. Stand in front of the tree with us. Who has the longest arms?"

"What kind of kinky shit are you guys into?" Callen asks.

I smack him and hiss when my hand hurts instead of his chest. "Asshole. I want a selfie with all of us. If the rags are going to gossip and guess, let's at least give them a good pic for a change."

Brynlee takes my phone from my hand. "It's called a timer, Linds. Give me a second."

She sets the phone up on a shelf and runs back over to us, and we all squeeze together.

"Everyone say *Fuck the press*."

A round of *Fuck the press!* gets called out as we all laugh, and I post it on my own social-media account for the first time in weeks.

First Christmas with my hot hockey-god husband and my family. #FoundFamily

That ought to shut them all up.

The Philly Press

KROYDON KRONICLES

PICTURE PERFECT

OMFG, peeps. Check out baby Kingston's most recent post linked below. Look at all the gorgeousness in one shot. I spy with my little eye . . . two pro athletes, a dancer, a cheerleader, and our favorite Olympian among this group of incredibly attractive social elites. This reporter would love to be the cream inside any of their cookies. Could it be that Hazey is finally debuting in society with this snap? #Hazey #KroydonKronicles

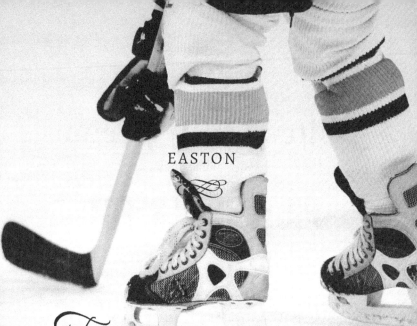

EASTON

Fitz always has the team meet for a nine a.m. skate the day of a game. We hit the ice light for half an hour, then have to be back at the Battleground Arena two hours before puck drop. Max Kingston gutted this place while I was in high school. It had been falling apart before he bought it. Always on the list of worst arenas to play in. But now... Now it rivals any new arena in the country.

We're moving slow this morning, but we don't typically move at anything close to even half speed before games. Today is a little different for me though, because tonight, we play the Vipers. Nobody wants to go up against his former teammates, but especially not while it's still so fresh.

"You got anything to add, Hayes?" Fitz asks after he calls us off the ice for the morning.

"Yeah. We've all watched the tape. You know these guys are good. You're better. You're more seasoned. They brought in a lot of new guys this year to try and stay under the salary cap and still keep their center happy. But they know me. They know my game. They know how I move. I've trained with them for ten years. I'm gonna try not to be predictable for them, but I could use your help tonight." Nobody wants to look like they need help from their teammates, but I need to make sure they're thinking this way if we're gonna win.

Jace nods and looks around at the team. "You hear that, guys? Our goalie is laying it on the line. Are we gonna let him down?"

Boone throws an arm around me. "Fuck no, we're not."

"Fuck no, we're not," Jace echoes. Guess we've come to an understanding after all.

He follows me out of the locker room after practice. "Hey, Easton. Wait up."

"Yeah, man," I slow down as we head for the door. "What's up?"

"You going with Lindy to meet with Sam today?"

"Yeah. I'm picking her up now, and we're going over together." I push through the door and stop in the parking lot. "Why?"

"Just glad she's doing it." He opens the door to his SUV and throws his bag inside. "She might not have needed that level of security ten years ago, but she needs it now. I appreciate you convincing her."

"I didn't convince her of anything. I told her I wanted her safe, and I wanted her in control. She needs to look at security as a tool that lets her live her life without being afraid of getting bombarded every time she leaves the house. I need her safe, man. But it's gotta be on her terms."

"You know I fucking hate when I'm wrong, Hayes."

"Yeah, man. Most of us do." I shoulder my bag, ready to get out of here and get to Lindy.

"Yeah, well it's worse when it's in front of the kid who could barely talk to me he was so starstruck the first time we met. I let Lindy down, and I didn't really worry about letting you down in the process. I should have. She's just always clouded my judgment. She's my baby sister, you know?"

"Listen, I wouldn't exactly say I was starstruck," I laugh. "But I guess I get it. I've always wanted to protect her. The idea of letting her down has haunted me for fucking years. But I'm not the one you need to worry about hurting her. I never will." I shove his shoulder, not sure what else to say. "I think she's planning on coming to the game tonight. Maddox said he'd come with her if she doesn't have security in place yet. Any chance you could try to smooth shit over a little with your family?"

"You're good for her, ya big fuck. You know that?"

"Yeah, man. It might have taken me a few too many years to accept that. But she's not getting rid of me. So I hope I'm good enough."

"None of us are good enough for the woman who falls in love with us, man. None of us. We just have to do our best to earn it every day."

Jace gets in his SUV, shuts the door, and waves as he pulls away.

None of us are good enough.

Guess it's a good goddamn thing hard work doesn't scare me.

Sam Beneventi's office isn't a place I expected to ever find myself.

I've watched my fair share of mob movies in my life. Read a few books. Heard a few stories. You can't avoid them. Especially when you live in Las Vegas. But none of that prepared me to sit in this office and ignore the fact that Lindy's brother-in-law, Maddox's dad, is the head of the Philly mob. From what I've heard over the years, he owns a shit-ton of

THE KEEPER

legitimate businesses too. But the Beneventi crime family has run this city for a century.

Sam's in his forties. But other than a little gray mixed in with his dark hair, he doesn't look much older than Maddox. Power rolls off him in waves. He's intimidating as fuck.

When he speaks, you listen.

"Explain to me what you want handled differently, Madeline, and I'll let you know if we can do it. I'm going to be honest with you. My team are professionals, and they make the final calls. You know that. If my men feel like they need to handle a situation a certain way to keep themselves and you safe, they're going to do what they have to. But they'll protect you at all costs."

Lindy sucks in a breath. She knows that fact all too well. One of Sam's guys, Marco, was her bodyguard on duty the night I took her home from a Kings game. I remember it like it was yesterday.

*S**he was fifteen years old and spent the whole day trying to flirt with me.*

Not that I didn't want to flirt back, but fuck. She was fifteen, and I was nineteen, about to turn twenty. Nearly five years may not seem like a lot later in life, but it felt like it back then. I kept reminding myself she was too young and off-limits.

I drove Kenzie and her to the Kings game that day, but Kenz was spending the night at Brynlee's, so it was just Lindy and me driving home, with her bodyguard, Marco, following in the car behind us.

I remember pulling into the driveway and turning off the car.

I remember talking about hockey and teasing her about figure skating. Teasing her about her partner, when deep down I was jealous because he was her age. That if something was going on between them, it was okay because he was the better guy for her. I'd never even told Pace that. Nobody knew how I felt because she was so fucking young.

She laughed at something I said, and it was the sweetest sound I'd heard in so fucking long. I was drafted into the NHL at seventeen. Everything was thrown at me after that. Women. Sex. Booze. Drugs. It was all there for the taking. There was no innocence in my life, not that there'd been a ton since my mom died.

I forgot myself for a second and cupped her cheek before I pulled my hand back and got out of the car. The look on her face was so fucking hurt. But fifteen kept flashing like a broken neon sign in my mind.

Marco walked ahead of us to the door, and she teased him as she handed him her key.

"You don't have to be so dramatic, Marco. It's my house. It's not like we don't have an alarm system you guys installed."

He ignored her teasing and walked through the door with Lindy following behind and me bringing up the rear. By the time I pulled the door shut behind me, Marco was dead on the floor, and a man I'd never seen before was standing over him with a gun hanging down at his side.

There was no sound.

I remember being so confused in that moment.

Not understanding there was a silencer on the gun.

Not having a clue what was going on until bright-red blood poured out of a hole in the center of Marco's head.

Lindy turned away, and a high-pitched scream I'll never forget ripped from her throat.

Holy fuck. He shot him.

"You two." *He jerked the gun toward us, and I moved Lindy behind me.* "Over there. Now. And don't try anything. I don't want you—either of you. I want your mom, little girl."

Lindy sobbed hysterically as my fight-or-flight instinct kicked in, and I had to push it down.

Holy shit. What was I supposed to do?

I sat us both down slowly on the couch.

Who the fuck was this guy? What did he want?

"Henry," she cried. "Why are you doing this?"

Henry pointed the gun at her. "Because Ashlyn was supposed to be mine, not his. I need to make her see that. That's where you come in. We're supposed to be a family. Not him."

He had to be talking about Brandon Dixon—Ashlyn's new man.

What the hell was I supposed to do with all this?

Henry walked around to the back of the couch laughing at something. Who knows what. He pressed the gun to the back of my head, and I'll never forget thinking I was gonna die.

"You're pretty useless in this whole scenario. Maybe I should get rid of you now."

"No," Lindy sobbed. Her crying escalated, and Henry hit her in the back of the head with the gun.

"Calm her down and you can live."

I wrapped my arms around her and pressed my lips to her ear. "Shh. You gotta calm down," I begged her loud enough so he could hear me. Then I whispered quietly, "I'm gonna get us out of this, princess. You gotta trust me."

Henry cocked the gun. "No whispering."

I held Lindy tight, trying to figure out what the hell to do as he pressed the gun to her head again. We sat there, waiting. Silent tears racked her shaking body.

I'm not sure how much time passed before Ashlyn walked in, followed by Brandon. She screamed as she stumbled over Marco.

"Not a sound, Ashlyn," Henry warned her.

"Henry, what are you doing?" Ashlyn took a step forward, but Brandon pulled her back against him, and I watched Brandon's eyes. He was a big guy back then. Still is, but then, he was the

center for the Philly Kings football team, and I remember thinking if we were gonna get out of this, he and I had to work together.

"Mom," Lindy cried out. *I lock my arm around her, keeping her in place.*

We were not fucking dying that day. I remember thinking that like a mantra over and over. Trying to convince myself.

"What are you doing, Ashlyn? That's the better question. Why are you living a lie?" *Henry asked her, completely unbothered by any of it. He was calm, like he hadn't just killed somebody's husband. Somebody's dad.*

Ashlyn tried to get Henry talking. I think she was trying to distract him. But it wasn't working. He was just getting more pissed.

"Step away from him, and I'll explain everything, my pet." *Henry motioned to Ashlyn with the gun, and Brandon's hold tightened.* "Let go of her." *The gun moved to Brandon, and I was sure Henry was about to shoot him.* "It's his fault. He ruined everything."

Ashlyn immediately moved away from Brandon toward the couch. "What did he ruin, Henry? I'm so confused."

"How are you confused?" *He waved his gun around, and Lindy's nails dug into my leg as this psychopath put the gun back to her head.*

"What's he doing here, Ashlyn? He shouldn't be here."

"He lives here," *she told him.*

"I'm supposed to live here. Not him," *Henry yelled back, and the sinking feeling came back tenfold. How are we getting out of this?*

Henry was screaming at her. Spit flew from his lips. "You saw me. You finally saw me. After all these years. Do you have any idea how long I waited for you? Any idea the lengths I've gone to so we could be a family?"

I tried making eye contact with Brandon, but his eyes were locked on Ashlyn.

If I couldn't get him to see me, we were all gonna die.

"I've loved you for so long, Ashlyn. Since your very first Nationals when you were fifteen. And you never saw me. Do you remember the way the seats would fill when you practiced during public hours? Do you know how many hours I sat in those stands, waiting for you to notice me? All those hours on the top bleacher, watching your every move. Every routine. Always hoping that would be the day you'd see me. You were so pretty. So graceful. I was there when you won your very first Nationals. I was there at the Olympics when you were robbed of the gold and that spoiled little bitch, Nina, threw a temper tantrum on the ice."

Henry waved his gun around the room, and the cracks in his calm started showing. This was going downhill fast. *"She ruined your chances, Ashlyn. She tainted you. It was all her fault that you were robbed of the gold. There was no way they were going to give it to you after the stunt she pulled. You'd worked so hard for the medal. It wasn't fair. So she had to go."*

The room became eerily quiet until he pulled back on Lindy's hair, and she cried out.

I fought everything inside myself to stay calm and focused on Brandon, not on killing this asshole for hurting Lindy and threatening our lives.

I refused to fucking die like that and pushed down my fear.

I forced myself to stay in control as Ashlyn kept Henry talking.

She knew what we needed. Now let's just hope Brandon was understanding me. "What do you mean, she had to go, Henry? What . . . what did you do?"

I tuned out Henry's answer and gave Brandon the slightest nod to see if he'd notice, and his eyes widened. Fuck. He saw it. I made promises to God if we got out of this, I'd do whatever it took to lead a good fucking life. And as this crazy fuck yanked on Lindy's hair again, he pressed the gun tighter to the back of her head.

Ashlyn saw it and forced her way around Brandon, still arguing with Henry. Keeping his focus on her, instead of Lindy.

Henry lowered his gun, then pointed it at Brandon and Ashlyn.

"We're supposed to have a life together, Ashlyn. I was even going to forgive you for her."

I just had to wait for my time.

It was coming. I knew it had to be coming. We weren't dying like this.

Not there. Not that day.

"We still can, Henry. You and me. Just let Madeline, Easton, and Brandon go, then I'll go anywhere you want. As far away as you want." Ashlyn took a tentative step closer, and Henry moved.

I remember thinking—That's it, Ashlyn. Get him to move.

"Anywhere, Henry. We can start our lives together anywhere. But you've got to let them go."

Henry swung the gun toward Brandon. "He'll never let you go."

That was it. That was my chance. Thank fuck, Brandon was used to reading silent signals on the football field and knew innately what I was thinking.

I nodded at him, and he threw Ashlyn down on the floor at the same time I pulled Lindy down in front of me and out of the line of fire.

In a lightning-fast move, I twisted my body and grabbed Henry's wrist with both hands. I was trying to control the gun.

Looking back, it happened so fucking fast, but it felt like I was slogging through quicksand back then.

I yanked Henry forward and ripped him off his feet, praying the gun wouldn't go off and kill anybody.

This was it.

Our only chance.

I've never been as scared in my entire fucking life as I was when the gun went off. I didn't know if it hit anyone until later, when someone told me it went into the wall.

Brandon hurtled his body over Lindy and me, like the couch was a fucking springboard.

He tackled Henry to the floor behind us, knocking over the fucking couch, with Lindy and me both still on it, in the process. We all fell to the floor as momentum carried us.

I threw Lindy at Ashlyn and turned to help Brandon, who had his hands around Henry's throat and was slamming his head against the floor over and over.

Blood pooled beneath the back of Henry's head as his face turned a dark purple.

Fuck.

He's gonna kill him.

I have no fucking clue, even all the years later, how or why I pulled Brandon off. But I drug him back from Henry's motionless body, lying limp on the floor in his own blood.

I wrapped my arms around him from behind and looked up when Sam's cousin, Dean Beneventi, walked in front of us and grabbed Brandon's face.

I'll never forget the moment.

"You gotta stop," Dean yelled. "You're gonna fucking kill him. And as much as you want to, you can't. Your family needs you. Go to them. I'll handle this."

Brandon pulled away and screamed at Dean, and I turned to look at Lindy, who was sobbing and shaking in Ashlyn's arms. I was frozen in my spot for the first time all night. I couldn't move.

We almost died.

She almost died.

Movement drug me out of my moment as Brandon ran to Lindy and Ashlyn, and Dean movds next to me. "You okay, kid?"

Was I okay?

No.

Nothing was okay.

I heard Lindy sob from across the room. "Marco. He . . . he shot Marco." Then she pushed away from them, frantic. "Where's Easton?"

"Right here," I told her, and she climbed over Brandon and threw her arms around me.

"I thought he was going to kill us," she cried.

"I was never gonna let that happen, princess," I told her. I held her so fucking close, and in some ways, never let go.

"*E*aston . . ." Lindy lays her hand on mine, bringing me out of a past I try really fucking hard not to revisit. "You okay?"

I look at my beautiful wife, alive and happy, and kiss the top of her head. "Yeah, princess. I've got everything I need."

LINDY

EVERLY

I wasn't born to work a 9-5. I was born to swim in warm turquoise water, drink cocktails out of a coconut, and chill in a villa by the ocean with an attractive man who doesn't speak English.

BRYNLEE

Why wouldn't he speak English?

KENZIE

Because that sounds like some tiny little far-off island. He probably has some sexy foreign accent.

BRYNLEE

Ohh. Yes! I thought she meant he didn't speak. Like at all.

EVERLY

That is what I meant.

KENZIE

Gonna need you to explain that one to me.

GRACIE

Do you know her at all? She's saying she wants to use him for sex but doesn't want to have to talk to him after.

KENZIE

Oh.

BRYNLEE

Oh.

LINDY

Thank God Gracie can translate Everly for us.

GRACIE

How'd your meeting with Sam go?

EVERLY

Yeah. Are we getting Chuck back?

LINDY

No. Charles has already been reassigned. The new main guy's name is Crew. He actually trains with madman at your dad's gym, Bryn. There are two others who will cycle in when I need them, but Crew's my new Charles.

KENZIE

Is he hot?

GRACIE

Is he single?

EVERLY

Is he straight?

BRYNLEE

Is he going to report everything we do back to my dad?

LINDY

> Yes. I don't know. I think so. And no. I made him sign a nondisclosure agreement. He'll be with me at the game tonight. Anybody wanna come?

BRYNLEE

Already gonna be there.

GRACIE

Can't. Nutcracker opens in three days.

KENZIE

I'm with you.

EVERLY

Me too.

LINDY

> Love you ladies.

"Hell-o lovah . . ." Everly whispers as she walks into my condo. "Seriously, you didn't say that man was panty-melting hot. How is Easton okay with *that man* protecting you? Isn't he worried about your body and his body . . . you know?"

"You're crazy. You realize that, right?"

Kenzie tosses me my coat and looks at her watch. "She knows. We all know. Now let's move it, ladies. I can finally catch one of my brother's games, and I don't want to be late."

"Me either. Have you seen the way those men stretch on the ice? I swear my TikTok feed is filled with hot hockey boys lately. It's like the sports gods know how over football players I am by this point in the season."

I slip my arms in my coat and grab my purse. "It's a good

thing you're not allowed to date any of the Kings players then, isn't it?"

"Who's dating one of my teammates?" Callen asks, having caught the end of the discussion when he and Maddox walked through the door.

Evie stomps her feet. "Why are dumb and dumber here?"

"Play nice," Kenzie tells her as she takes my hand. "Now let's go."

My family purchased the Philadelphia Revolution from my oldest brother Max's wife's family years ago. And one of the first things he did was institute the same rule here that Scarlet has at the Kings stadium. We have two box suites. One is for the family to enjoy themselves. The other is for VIP guests, who we're expected to mingle with.

If you're one of my brothers, sisters or me and you're at a game, you're expected to pop into the VIP suite at some point and make nice. But as we pass it tonight, I can't help but wonder if I wouldn't be better off staying in there with my friends instead of going into the family suite.

I know Max and his wife, Daphne, will be here because I spoke with him earlier to make sure there was room. Most of my nieces and nephews are teenagers and rarely come to the hockey games. It's hard to make them all when there are so many more hockey games than there are football games. Home football games get the majority of us there, most of the time.

"Easton told me Becket and Jules were coming with Blaise, which means there's a good chance Lenny will be here too," I tell Kenzie. "Am I awful that I'm hoping Mom and Brandon skip this one?"

"No, you're not awful. But you have to face her soon, Linds. It's almost Christmas. You don't want to let this ruin your holiday."

I square my shoulders and walk into the suite, taking a quick glance around.

I see Max and Daphne. Jules and Becks. I was right—Lenny and her husband, Bash, are sitting with them. And India is here with the kids. Okay, I can handle this.

"Lindy," Becks is the first to see me, and the smile on his face does me in. He wraps me up in a big-brother bear hug, and it's like I can finally breathe. Jesus. I forgot how much I miss my family.

"Hey, Becket." I squeeze him back and kiss Jules on the cheek when she joins us.

"So, kid. Does this mean you're my daughter-in-law and my sister?" Becks jokes, and Jules smacks his chest.

"Don't let anyone hear you saying that, Becket. It makes us sound like the Clampetts."

"I'm just teasing her," Becks chuckles and moves on to Kenzie, not making a big deal out of anything. Okay. New favorite brother.

Everly walks over before I even take my jacket off and hands me a glass of wine. "Here," she whispers. "Take the edge off. You're practically vibrating with nerves."

She taps her glass to mine. "Salut."

"Salut."

"You know, according to the Kroydon Kronicles, you shouldn't be drinking." Lenny eyes me as she and Max make their way over to me.

Max shakes his head. "Like she'd be the first of us to be pregnant before she got married, Eleanor."

His wife, Daphne, shakes her head. "Whatever, Maximus.

Like you were the one who had to carry around a bowling ball for nine months."

I sip my wine while Everly laughs quietly. "I always forget how crazy your family is."

"Uh . . . guys?" I hold my glass up. "I'm not pregnant. Promise."

Lenny leans in and kisses me. "Good. Enjoy all the fun you get to have with him before you have a baby. Now is when you get to screw on every surface of your home without having to worry about waking the kids up."

Max gags and walks away, while Everly taps her wine glass to Lenny's beer bottle.

"You know, Everly, there's always this weird line where I want to say these things to my little sister but somehow still feel weird about saying them in front of you."

Evie and Gracie grew up with Lenny basically as a member of their family. Their parents are close. Lenny's husband, Bash, is Evie's godfather.

"It's not like I'm a blushing virgin, Len," Everly swallows her wine, and Len scrunches her face up.

"So." Len turns back to me. "Have you talked to your mom yet?"

Jules and Kenzie turn toward our conversation, and I suddenly feel bad. I'm not happy with my mom, but it feels weird talking about her with everyone.

"No. Not yet," I admit.

"You need to call her, Lindy," Len lectures gently.

"She hasn't called me either," I defend.

"Sweetie." Jules takes my hand in hers. "If you want to be treated like an adult, you've got to act like one. And sometimes that's realizing this is your mother, and you owe her your life. Literally."

"If I say I'll call her, will you all let me watch my husband warm up?"

"She called him her husband," Jules snickers to Lenny. The two of them have been best friends forever. And in moments like these, I see it.

"She did. Our little girl grew up," Lenny adds.

"You guys . . . Ugh." I walk away, laughing, and move to the glass to find Easton. He's skating lines in front of his goal. And damn, my husband is hot.

I promised Easton I wouldn't wait for him after the game. He wanted me to go home with Crew and my friends, since we're subtly trying to avoid the media until some of this firestorm settles down. That means I've got time to kill while I wait for him on the couch, snuggled up with Myrtle later that night.

I stare at my phone, trying to decide whether I want to call my mom or not.

I almost do it too. Until I chicken out and call Brandon instead.

"Hey, shortcake. I was hoping we'd hear from you soon."

Damn it. His voice wasn't supposed to make me cry. This is why I didn't FaceTime.

"Hey, Brandon. How's everything?" I ask and feel like a complete asshole.

"Gonna ask me how the weather is next, kid?"

I pull my chunky white-knit blanket around me and kinda wish I was sitting next to him right now. Brandon didn't come into my life as a stepfather until I was fifteen, and he's never treated me any differently than my little sister, Raven. "No. How is she?" I ask instead.

"There's a lot of shes in this family, Madeline." His voice is firm. Protective. He loves me, but there's no one in this world he would choose over my mother.

"Mom. How's Mom?"

"That's a loaded question. Might be better off if you ask her. How are you?"

"I'm happy. Everything in my life is great, if I ignore the fact that my mother wants me to stay a child she can control forever," I grumble, frustrated. "And I'm pretty sure I sounded like a petulant child the way I said that."

"Kid. Your mom is hurt. She loves you, and you threw her a Hail Mary, expecting her to catch a pass she didn't know was coming. She dropped the ball, but that doesn't mean it can't be recovered."

"There's my favorite football coach," I tease. He tried being an analyst for a season after he retired, but it wasn't for him. Now he's the offensive line coach for the Kings, and he's great at it. But . . . "Can you translate that to layman's terms, big guy?"

"You threw her for a loop, kid. She wasn't expecting you to come home married. She wasn't expecting you to end up on every gossip rag there is. We don't even know whether the news is true and we're going to be grandparents or not—"

"You're not," I confirm. "I mean, maybe one day. Eventually. Many, many, *many*, years from now. But not yet and not in nine months."

"Then you fired your security. She's been worried about you."

"I fired my security because he wasn't mine. He was theirs. The family hired him. I hired someone new today. He works for me and answers to me. I'm not trying to be irresponsible. I just need her to let me breathe a little." I run my hand over the soft fur of Myrtle's head and try to see this from their perspective, but it's just so damn hard.

"Any chance she's going to be around tomorrow night?" Maybe it's better if I just rip this Band-Aid off.

"Not tomorrow. She and Raven are going with Jules and Blaise to see some Christmas show tomorrow night. But you could come keep me company if you want to."

"Big guy, if I come and hang out with you before I talk to Mom, you'll be sleeping on the couch for a week." My tone is teasing, but I'm pretty sure I'm not wrong.

"Shortcake. Don't ever make Easton sleep on the couch. Argue about something if you have to, but put it to bed before you go to bed. Life's too short to be that angry."

"How'd you get so smart, Brandon?" He's always been this person—the one who gives the best advice.

"I'm surrounded by smart women. I pay attention. Now call your mother."

"I will." I yawn and slide my head down to the couch. "I love you, big guy."

"Love you more, kid. Good night."

"Good night."

He always used to tell me he loved me more. One day, I asked him why he always had to outdo me, and he laughed and told me, "I don't love you more than you love me, shortcake. I love you and Raven and your Mom more than anything else in the world. More than anyone else ever will."

It might have been true back then, but I think I found someone who loves me more now. A different kind of love. I guess that's how it's supposed to be.

EASTON

Fitz splits his practice. We're all on the ice at the same time, but half lift before practice, and half lift after practice. I'm a before guy, so when I get to the ice, I've already been going for a few hours. I smile at Crew. He's standing at the back of the tunnel leading out to the ice, keeping my girl safe.

It's been strange getting used to him always being around this past week, but I was glad he was there when I had two back-to-back away games over the weekend. We go over Lindy's schedule each night, and he adjusts his time accordingly. Switching off who's going to be on her, depending on what she plans on doing and when. "Hey, man," I shake his hand, then move around him to watch my wife skate for a few minutes before I head back to the locker room. I always get here early now, just so I can watch her.

Lindy is working with her baby skaters today. And when I say baby, I mean baby. These kids look like they should barely be able to walk, let alone skate. She sees me and waves, causing all the moms sitting in the stands to turn and look. Oops.

Once I'm changed, I come back out and find her skating by herself. She stops when she sees me and glides over. Her little pink skirt barely covers her ass, and the gray sweater she's got tied around herself doesn't look like it's doing anything to keep her warm. "Hey, hockey boy."

I grab her around the waist and skate her backward. "Hey, princess. How's your day going?"

She spins out from me and takes my hand in hers. "It's good. Better now." She pulls up and drops a quick kiss on my lips, then pushes off.

"What are you doing, princess?"

"Think you can catch me?" she taunts, and my dick gets instantly hard.

"Yeah, Hayes. Think you're faster?" Boone yells as he walks up to the glass.

"No way. I've got ten bucks that says my sister's faster than Hayes," Jace adds as he joins Boone. Before I know it, half my team, including Fitz, are arguing over who could win a skills competition. They decide Fitz is the judge, while Lindy and I stand there, staring at them.

"You're all delusional if you think I'm gonna do this," I tell them.

"What's wrong, hockey boy?" She skates a circle around me. "You scared of a figure skater?"

I shake my head and whisper, "I'm gonna make you pay for this, wife."

"Promise?" Her eyes light up with mischief, and all sorts of ways to make her pay come to mind.

"How much money you got collected over there?" I shout.

"Enough for drinks at the bar tonight," Smitty yells back.

"I'll make it worth it for you, Hayes," Fitz tells us. "You win, and practice is canceled."

The team cheers, but Lindy skates over to Fitz. "And what if I win?"

Jace moves closer. "If you win, they all have to help with your baby skaters for a week."

"Deal." She shakes Fitz's hand, and I have a bad feeling this isn't going to end well.

"What are the tests?" Lindy asks and crosses her arms over her chest, plumping up her boobs.

"Well, that's one way to win," I whisper in her ear. "Just distract me, why don't you?"

"Speed," one of the guys yells.

"Obviously," Lindy agrees. "Forward and backward."

"Slap shots," Boone adds.

"Fine." She nods her head, thinking. "Then Easton's got to do a spin."

"Okay. How hard can a spin be?" I ask, and the guys all grumble.

"Bring it, hockey boy. Let's go." She kisses my cheek and skates down to my net. "I don't have all day, husband."

My chest shakes with laughter.

I fucking love her.

*T*he guys all cheer and get the fuck out of dodge when we're done with our little competition, while my wife pouts over losing. Jace doesn't help matters much.

"You won a fucking gold medal. You couldn't skate faster than this giant? Come on, kid. You were supposed to kick his ass."

"Go home and enjoy your free afternoon, Kingston," Fitz tells Jace after everyone files out before he turns toward Lindy. "You're a beautiful skater, my dear. But Hayes's legs are twice as long as yours. It wasn't a fair fight. A gentleman would have let you win."

"She would have kicked my ass for a week, if I let her win," I laugh, and my wife wraps her arms around my waist and kisses my cheek.

"Aww. You really do know me."

Fitz shakes his head. "Gotta love a good sport. See you on the plane tomorrow, Hayes."

"Thanks. See ya, Coach." I watch him leave and look around. "Looks like it's just you and me, princess."

"And Crew," she reminds me.

"Hey, Crew," I yell.

"Yeah?"

"I've got her from here. Take the rest of the day." I don't bother waiting for an answer before I lift her off her feet and skate around the rink. She weighs nothing in my arms as she leans back in some fancy move and bends her legs and back, so her head looks like it's going to touch the skate.

When she pulls her head back up, I kiss her like I've been wanting to do all day. She tastes like chocolate and cinnamon, and I'm fucking starving for her. "You were so close, baby. You almost had me."

"No, I didn't. Don't patronize me." She licks my lip, then bites down. "Now what's your prize, hockey boy?"

"You're my prize."

"I am, aren't I?" her lips curve into a seductive smile. "If you take me into the locker room, I'll show you just how much of a prize I can be, husband."

We walk back into the empty locker room before I set her on her feet, and we both take off our skates. "What do you have in mind, princess?"

She takes my hand in hers and moves us into the back corner of the room, making sure no one else is here, then grabs a towel from a shelf. "Can you be quiet, Easton?"

"Madeline. Anyone could come in here," I warn.

A pretty pink rushes to her cheeks. "Then I guess I better make you come fast." She tosses the towel on the floor, then drops to her knees in front of me.

Logically, I know this is a bad fucking move. But when she looks up at me through those dark fucking lashes, I'm no longer thinking with my head. *Not that head, at least.* "You look so good on your knees, baby."

I wrap her ponytail around my fist and tug, enjoying the quiet moan she gives me in return.

Lindy pulls my sweats and boxers down to my knees, and her stormy eyes dance with excitement. "I love your cock, E."

"Then I guess you better show me how much, wife."

She licks her lips and fists my dick. And I swear, the tiny moan slipping past her lips as she swirls her tongue over the tip of my dick is the sexiest fucking sound I've ever heard.

She drags her tongue down to the base of my cock and cups my balls.

And *fucking hell*.

Blood roars in my ears.

My wife knows exactly what she's doing to me because she looks up at me with such an innocent smile before wrapping her pouty lips around my cock. Taking me deeper until she swallows me down her throat and fucking hums.

"Fuuuuck . . ." I pull her hair until she stops. "Baby, you gotta stop or I'm gonna come."

"Good." And this time when she takes me down the back of her throat, I don't stop.

When we get in the SUV, Lindy's phone connects with the Bluetooth, then it immediately rings through the car.

The system announces, *Mom calling*, as I pull out of the parking lot. "You gonna answer her?"

"I guess . . ." She looks at her phone for a second.

"Princess, you've got to talk to her. Christmas is in a week."

"Fine."

I hit the button, and Lindy throws me a look. "Hey, Mom."

"Hi, Madeline. Why do you sound far away?"

"I'm in the car with Easton. You're on Bluetooth."

I link my fingers with her and squeeze. "Hey, Ashlyn."

"Hi, Easton. Are you guys going out or going home?" she asks, and Lindy looks almost panicked.

"We're just leaving the rink to go home and grab dinner," I tell her, then wait as the beat of silence lasts a few seconds too long.

"Why don't you come over here for dinner? Your sister is spending the night at Hudson's house, so it's just Brandon and me. We haven't seen you in weeks. It would be nice to see you."

"Thanks, Mom, but we—"

I cut my wife off. Let her be pissed. She'll thank me eventually. "That sounds great, Ashlyn. Want us to pick up anything?"

"No. We're calling in an order now. We'll see you in a few minutes."

"See you in a few minutes, Mom." She disconnects the call and drops her head against the seat. "Really, E? I just sucked your dick. Aren't you supposed to be in a good mood?"

"I *am* in a good mood, baby. I want you to fix things with

your mom, and you've come up with every excuse imaginable to avoid her for weeks. Weeks, Lindy. Time's up. Christmas is a week away. You don't want to miss that with your family."

She turns her head and watches me as I navigate the few streets between the practice facility and her mom's house. "I hate when you're right, you know that?"

"I do, baby. But I'm doing this for you."

She takes a deep breath and sighs. "I still hate it. But I love you for caring about my relationship with my mom."

"I'll always care about your family, princess. They're mostly annoying as fuck, but they matter to you, so they matter to me too. Even Jace." We pull into her mom's driveway, and I walk around the car and open the door. "Now, come on. The sooner you make up with your mom, the sooner I can take you home and give you your prize."

"My prize?"

I pull her close to me. "Hey, I'm an equal opportunity competitor. You had better jumps and spins than me. Pretty sure that gets you a few orgasms."

"I love the way you think. Let's get this over with."

LINDY

We stop at the door and knock before walking in. Mom's rule has always been I'm supposed to walk in because it's still my house. But still, it doesn't feel right. "Hello . . ." I call out, then take Easton's hand in mine. "They're probably in the kitchen."

"We're in the kitchen," Mom yells back.

I turn to E and shrug. "Told ya."

We moved into this house after *that night*. I loved our old house. It was sandwiched between two of my brothers' houses. But even with that, none of us wanted to step foot in there again. This was the house Mom and Brandon bought together. It's where we became a family. Before this house, it was just Mom and me.

"Hey, shortcake." Brandon drops a kiss on my head, then offers Easton his hand. "Hey, Easton. Good to see you."

"You too."

Brandon smacks Easton's back. "Have I shown you the new sound system we installed in the gym downstairs?"

"Uh, no." E looks confused.

"He's trying to take you away to force a little alone time

on Mom and me," I groan and see right through my stepfather.

A smile breaks out on Easton's face, and he slaps Brandon's back. "You know what? I'd love to see that sound system. Lead the way."

He looks back at me and winks.

I mouth back, *I hate you,* then make my way down to the kitchen, where my mom's making a salad. She was barely my age when she had me after a completely fucked up life. It was just us for so long before Brandon came along. We had my brothers and sisters, but Mom and I were a team. Tears burn the back of my eyes as I walk in.

"Hey, Mom."

She puts her knife down and pushes away the cutting board full of tomatoes. "Oh, sweetheart. Don't cry. If you cry, I'll cry."

We both move at the same time and wrap our arms around each other. "I'm sorry I didn't ask you what you wanted and how you felt, Lindy. I was so caught off guard and mad about the way everything happened. I just dug my heels in, and the harder you fought, the harder I pulled."

"You shouldn't have had to find out the way you did, and for that, I'm so sorry. But Mom, I'm not sorry for marrying Easton. I love him," I tell her and take a step back.

"That should have been what I asked you that day." She cups my face like she used to when I was little, her eyes filling with her own tears. "And does he love you, Madeline?"

"He does." I smile, thinking about just how much. "He's been bugging me to fix things with you from the beginning. He kept saying he'd give anything to see his mom one more time, and I was wasting time with you." My heart tugs, just from saying that out loud. "He's such a good man."

"And where is he? And Brandon? Are they hiding?" she asks, then pours me a glass of wine.

"Brandon's showing Easton something in the gym. I think he wanted to give us some space." I sit down and sip my wine. "I hired a new security detail."

"I heard," she muses as she finishes dicing her tomatoes while I watch.

"Did Sam tell you?" I ask, already annoyed. He wasn't supposed to say anything.

"Sweetheart, how many Revolution games has that man been at with you? Your family spreads gossip faster than a teenage girl. Now, we need to talk about a few things." She adds her tomatoes to a big salad, then sets it aside to eat when dinner gets here.

"We do," I agree.

"You said some things at Sweet Temptations I've never heard you say before, and I need you to talk to me. Did you feel like you had to skate for me, Madeline?" There's a shakiness in her voice, and it makes me feel like shit because I put it there.

"No, Mom. I never felt like you forced me into skating. I skated because I loved it. I still do. But you said I'm floundering because I'm not sure what I want to do right now, and that stung. I don't think I'm floundering."

"It's just so not like you to not have a direction. And then you went to Vegas, and Everly posted those pictures, which looked so bad. And we had to discover you were married on social media. It just all hit so hard."

"I like to think I'm transitioning right now. I spent my entire life training for the Olympics. I missed so many things. And now, I don't have to train any more. Now, I skate because I want to with no pressure. I teach the kids because it's fun. I've got more money than I could ever spend, and I don't want to miss anything else, so I'm not in a rush to jump into a job that isn't what I want. And I'm lucky enough that I can take my time figuring out what it is I want.

But the one thing I'm absolutely sure of is Easton. He's what I want."

I reach across the counter and rest my hand on hers, knowing this is going to be hard for her to hear. "I've never been happier than I've been with him, but no one in this family was happy for me. It hurt. And even worse, you all tried to make Easton into the bad guy. You were all supposed to love him, and you turned on him."

"Madeline." Mom pulls away. "We didn't turn on him. We needed you to help us understand. And I don't think any of us did a good enough job of meeting you in the middle. I *will* say that Lenny and Jules were on your side. They stayed quiet while we figured out what was going on, but they never stopped giving me grief. Becket too. Don't worry about this family turning their backs on that man. We all love him. But we were as mad at him as we were with you."

"He wants to marry me again. In front of everyone."

"Oh." Mom is careful not to give away her thoughts. "And what do *you* want?"

"I don't want some big, stuffy thing. That's so not me," I tell her honestly. "We're already married. I don't really see the point."

"The point, princess"—Easton wraps and arm around me from behind, startling me—"is for us to promise to love each other in front of the most important people in our lives. Juliette pointed out to me that the people who love us should get to celebrate with us." He looks over at my mom. "Hey, Ashlyn."

Brandon walks in with the takeout. "Are we eating in here or the dining room?"

"Here," Mom and I both say at the same time.

"So? Come on, princess. Don't make me beg." Easton drops down on one knee, and I try to pull him back up.

"What are you doing?"

"Madeline Kingston Hayes. I have loved you for what feels like my whole life, and I promise to love you until we're old and gray. Will you marry me? Again?" he adds, and I laugh.

"Get up, hockey boy." I pull him up and press my lips to his. "If you really want to do this again, I'll marry you."

He lifts me off my feet and holds me to him. "Love you, baby."

"I love you too, you big goof. Now put me down."

He drops me back to my feet, and my mom and Brandon hug us both. "I've always wanted to plan your wedding," she whispers.

"Small, Mom. Just family and a few friends," I tell her.

Brandon clears his throat. "Your family is bigger than the average person's wedding, shortcake."

"Don't remind me."

Easton

I pull Brandon to the side while Lindy's hugging her mom goodbye. "The only thing I regret about marrying Lindy in Vegas is not asking you for your permission first."

"She's her own woman. We both know that. You don't need my permission, and you've had my respect for a long time. There's no one in this world I'd rather see her spend her life with, Easton. Be good to each other. And be prepared —because you may think you know what being a part of this family means, but you don't have a clue until you marry one of them." He pats my back. "Don't say I didn't warn you."

The snow falls around us as we walk to the SUV, and

when I open her door, Lindy stops and kisses me. "It's like we're in the middle of a snow globe."

"I guess it does, now get in before you get cold."

"Thank you for making me come here tonight, E. You were right. I needed to do that. I know how much you miss your mom, but I've got to believe she's watching over you, and she's so incredibly proud of the man you are. I know I am."

I wrap my hand around her head and press my lips to hers. "I fucking love you, Lindy."

"Good. Then let's get home so you can show me just how much."

"Deal." She gets in the car, and I close her door and round the front hood. When I look behind us, there's a man in a sedan one house down. He's sitting in the car with the headlights off, and something about it feels wrong.

I turn our car on and look in the rearview, but he doesn't move.

Maybe I'm overreacting. But something just feels *off*.

"You buckled in, princess?"

Lindy looks at me funny. "Yeah, why?"

"I think that's a paparazzi behind us. Just being careful."

"Ugh, when are they going to stop following us? We're boring."

"You're never boring, baby."

I pull onto the street and watch to see if the sedan follows.

Thankfully, he doesn't.

Nothing like overreacting.

Stupid fucking tabloids.

Once we pull through the intersection outside of Ashlyn's neighborhood, a motorcycle flies up next to us—in the fucking snow—and the guy pulls out his camera.

"What the fuck?" Lindy gasps in shock.

"Ignore him. We're fine," I tell her, even though I don't like how close this guy is getting to us.

We pull onto Main Street, and headlights flash behind us.

It looks like the sedan from Ashlyn's neighborhood is back, and he's coming toward us at a pretty high speed, considering the snow that's already fallen tonight. "Is that fucker taking pictures too?" I shout, and Lindy turns to look, just as the motorcycle slides on the ice and veers in front of us.

I slam on my breaks to avoid hitting him, but it's too late.

He runs into us at my front corner. The bike slides across the hood of our SUV, and the guy collides violently against our windshield, just as the sedan slams into us from behind, sending us spinning into mass chaos.

Metal crunches, and time stops as I realize I have no control over what's happening.

"Baby." I look over at Lindy as our car comes to a stop in the middle of the road, and she screams.

I turn my head and am blinded by the oncoming traffic. Headed right toward us.

In a last attempt, I throw my arm across Lindy, helpless to stop what's happening. I hear a car lay on its horn and see it barreling down on us, trying to break. But I know he won't be able to stop in time.

Glass shatters, and the impact feels like an explosion as the front of the SUV crumbles.

The airbags explode, and the last thing I hear is my wife's scream before the silence is deafening.

EASTON

I wake up, disoriented and unsure of where I am before everything suddenly comes hurtling back to me.

The accident.

Lindy.

I bolt up and ignore the pain of whatever just ripped out of my skin. "Lindy," I call out, and Juliette and Becket come into view. "Where's Lindy. I need my wife."

Jules runs a hand over my face. "You need to calm down, Easton. You just ripped out your IV."

"Where's my wife?" I ask again, frantic. "Lindy . . ." I yell.

Becks grabs my hand.

The one that's not splinted.

What the fuck?

"We need you to calm down for a second and hear us, Easton. Take a breath." Becks refuses to let go of me when I try to wrench free. "Breathe, kid. We need you to calm down for Lindy."

His words break through to me.

"Why?" I look at him, fucking terrified. "Tell me she's alive, Becks. I need her to be alive. She has to be alive."

My eyes burn as tears I haven't cried since my mom died gather in my eyes. "Please, Becket. Please tell me she's alive."

"She's alive. She's in surgery now," he tells me, and I rip out what's left of the tubes and needles attached to my body.

"Easton." Juliette stands in front of me. "Stop. You're hurting yourself."

A nurse runs in as a machine beeps from somewhere in the room.

I look around like a caged fucking animal.

"Mr. Hayes. You have to sit back down." She turns away and calls out, "Someone get me help."

"Get out of my way and take me to my wife." I try to push past her, not giving a shit that she's a woman. She's keeping me from Lindy.

"You're not going to help your wife if you hurt yourself, son. Sit down. Let me take care of you for ten minutes, and we'll wheel you down to the private waiting room where the rest of your family is. I'm not going to keep you from her, but I'm not going to let you bleed all over the hospital either." She grabs a bandage and puts pressure on my unsplinted arm where my IV ripped out.

I lean back against the bed as the room starts to spin.

"Sit down, Mr. Hayes."

Becket moves around me and eases me down on the bed. "What's happening, Becks? Why is she in surgery?"

Becket looks to the nurse, whose badge hangs from her pocket saying her name is Helen.

She answers for Becket, "Your wife has internal bleeding. They need to find the cause of the bleed and stop it."

"You'll take me to her?"

"Just let me fix you up first. Now sit there and don't move. I'm going to get you some scrubs to put on."

Shit. I hadn't even realized I was in a hospital gown.

The nurse leaves the room, and I close my eyes. "What happened?"

"You were in a car accident," Jules tells me. "It was bad, but you had front and side airbags. They helped. It could have been so much worse." Jules wipes her eyes as she cries, and a guy in navy-blue scrubs walks into the room.

"Mr. Hayes?"

"Yes." My body locks down, preparing for the worst.

"I'm Dr. Midori, your orthopedic surgeon. You have a distal radius fracture in your right forearm, most likely from bracing during your accident. We were able to go in and repair it. You'll be in a splint for two weeks, while the swelling goes down, then in a cast for another four weeks. Once the cast is off, you'll be able to work with your team's physical therapist to get you back on the ice. That should take another four to six weeks. So all in all, you're looking at about three months before your back on the ice."

"Doc, I don't give a fuck about my arm. I need to get to my wife," I tell him, ready to crawl out of this room if I have to.

Nurse Helen comes back into the room with scrubs and a wheelchair.

"Helen, do you know where Miss Kingston is?" the doctor asks her.

"Hayes," I whisper. "She's Mrs. Hayes now. She changed her name last week."

Dr. Midori nods. "Do you know where Mrs. Hayes is?"

"I do. If you'd all give Mr. Hayes and me the room, I'll help him get changed and wheel him down to the surgical floor."

Becket clears his throat. "Is it okay if I help him instead?"

"That's fine, but don't let him fall. I'll be right outside this door." Helen walks out, and Dr. Midori stares at me.

"You aren't leaving the hospital, right? We'd like to monitor you overnight."

"If you can do it from my wife's bedside, then go for it. But I swear to God, doc. If you don't get out of my way and let me get to my wife, I'll go right through you," I warn him.

"We'll make sure he doesn't go anywhere, doctor," Jules tells him. "Thank you so much for taking care of him."

The doctor walks out of the room, and I wait for Jules to leave too.

"Not a chance, Easton Hayes. I'm not letting you out of my sight. I'm going to sit right here and keep my eyes focused on this wall while you get changed."

"Don't fight with her," Becket tells me. "The way you feel about getting to Lindy is how Jules and Kenzie have felt all night about you. Let her stay if it makes her happy, E."

I nod and kick my legs out so Becks can pull my pants up because if I bend over to do it myself, I have a pretty good feeling I'll fall the fuck over. My head is spinning from the drugs or the anesthesia or the accident. Pick one. It could be any of them.

He helps me get the shirt on, then Jules calls for Helen and the wheelchair. "I'm really not supposed to let you out of my sight, Mr. Hayes."

"You know where I'm gonna be, Helen."

She moves behind the wheelchair. "I do. And I'm going to take you there."

The hospital is quiet, with the lights off in most of the patient rooms. But once we get down to the surgical floor, there's no sign of the time. No way to tell that it's the middle of the night. Helen wheels me into a private room with the Kingstons, and Kenzie runs to me.

Becks stops her before she can launch herself at me. "Don't, Kenz. Don't hug him. Give him a minute. He's covered in bruises and has a fractured arm."

"I'm okay. How's Lindy?"

Jace comes over to me then. "She's still in surgery. They're supposed to come out here and update us once they locate the source of the bleeding, but we haven't heard anything yet."

I stand carefully, and Jace holds out his arm for me to grab onto as I slowly make my way over to where Ashlyn and Brandon sit, surrounded by the Kingstons. Scarlet rises from the seat next to her and touches my chest. "I'm glad you're okay, Easton."

I carefully squat down in front of Ashlyn, and she takes my hand in hers. "I'm so sorry," I say as tears fill my eyes. "I couldn't stop it. It's my fault. I couldn't save her."

"Easton," Ashlyn sobs. "There was an officer at the corner of the street. He saw it all. You couldn't have controlled what happened. The paparazzi caused the accident. The man on the motorcycle had a long-lens camera with him. He died for a stupid picture." Her voice shakes. "And Madeline—" She breaks off on a sob, and Brandon pulls her to him.

"Come on, Easton." Becket moves next to me and helps me into the chair next to Ashlyn. "You've got to take it easy. They'll come out and tell us what's going on soon."

*S*oon doesn't come for three more hours.

And when it does, you could hear a pin drop in the room.

A man and a woman, both dressed in dark-blue scrubs and surgical caps walk into the room. "Mr. Hayes?" the woman calls, and Juliette points them my way.

"I'm her husband," I say, feeling Ashlyn take my good hand in hers. "And this is her mother."

"Are you okay, Mr. Hayes?" the woman asks, and the room whirls around me. "I'm fine. Tell me about my wife."

The male surgeon answers, "Your wife is a fighter. Her seatbelt saved her life, but it also caused a splenic laceration. Once we located the bleed, we did everything we could to save her spleen but were unable to. She's out of surgery now and in recovery."

"What exactly are you saying?" I ask, wanting to make sure I'm understanding him.

"We removed your wife's spleen. She'll have to stay here for a few days, so we can monitor her recovery, and she'll have to take it easy while she recovers for the next four to six weeks. She's a lucky woman. This could have been much worse."

"Can I see her?" I ask, ignoring everyone around me.

"I'd normally tell you to wait until she's brought to her room, but you're not looking too good, Mr. Hayes. How about you let us take you back to your wife and we can look you over too?"

"Take care of my wife, and I'll be fine, doc. Just make sure she's okay."

Becket helps me back into the wheelchair, and Ashlyn calls out for me.

I stop and look at her.

I should feel guilty because I didn't offer to let her go back there to see her daughter, but I can't.

I need Lindy.

I need to feel her breathe.

"Take care of my baby, Easton."

I nod, feeling like I already failed, but I don't say anything as the doctor moves behind me and wheels me into Lindy's room.

The room is cold and quiet. The hum of the machines, the only sound.

The doctor wheels me over to her bed, and I rest my head against her arm as a nurse comes in and checks Lindy over. Then she looks at me, and I shake my head. She looks like she's going to fight with me but changes her mind and leaves us alone.

I press my lips against Lindy's limp hand. "I'm so sorry, baby. So sorry I couldn't save you. Please be okay. I can't do this without you." I drop my head down and do something I haven't done in fucking years.

I pray.

Lindy

The hum from the overhead lighting is the first thing I notice when I wake up.

The pain is the next thing.

"Lindy." My mom's voice pulls me further from the fog, and I open my eyes and try to focus.

"Mom." I find her next to my bed, with Brandon behind her.

I'm in a hospital room.

The pieces of a fuzzy puzzle start slowly falling into place, and I remember the accident.

I remember being wheeled into the hospital and told they needed to take me into surgery.

"Easton?" I ask, and my mom points to the other side of my bed, where my husband's head is laying on top of my hand. His arm is splinted, and his hair is a tangled mess.

"He hasn't left your side since they wheeled you out of surgery last night. His doctors wanted him to go back to his room, but he refused. He followed you from the recovery room to this room once you were admitted and hasn't moved since."

My eyes fly open, and I lift my numb hand and run it over his hair. "Hockey boy," I whisper, my throat dry and sore.

He doesn't move.

"I don't want to wake him up," I tell Mom and Brandon.

"Oh, sweetheart. That man was ready to take on anyone who got in his way to get to you. He was barely out of surgery when he got himself down to the surgical floor to wait with us. Wake him up and show him you're okay. That's the best thing you can do for him. We'll go find a nurse and tell her you're up."

Mom leans down and kisses my head. "I love you, Madeline. You're never allowed to do this to me again, got it?"

"I'll see what I can do, Mom."

She and Brandon close the door behind them, and I run my hand over Easton's head. "Wake up, husband," I call to him, and he mumbles something.

"Easton . . . I need you. Please wake up for me."

He moves his head and looks up at me. "Baby . . ." It takes a minute for him to focus, and then his hazel eyes transform from brown to golden-green, and he smiles and pushes to his feet. Immediately, his lips press against mine, and we just breathe each other in.

"Oh my God, baby. I thought I lost you."

I slowly reach up and cup his face in my hands. "You couldn't lose me. You saved me."

"I didn't save you. If I did, you'd still have a spleen and wouldn't be waking up in a hospital bed, Lindy." The utter heartbreak in his voice guts me.

"Easton, you did everything you could to stop what happened. We can't control other people or Mother Nature. And even then, when that car was coming right for us, you threw your arm in front of me." I gently touch his splint, and he winces in pain. "In my dreams, you always save me, E. And you save me in my reality too. You always do."

"I love you so much, princess. Old and gray, remember? You can't ever force me to live on this Earth without you."

He lowers the rail of the bed.

"What are you doing, E?"

He gently lies down next to me. "They've been trying to get me to lie down for hours. I'm following the doctor's orders."

"They're going to make you move," I tell him as I carefully rest my head on his chest.

"Let them try, baby. Let them try."

"I love you, Easton. Only you. Only ever you."

"Only ever us, baby. It was only ever us."

LINDY

"I hate that I'm wearing flats with this gown." I turn around and look at myself from behind in the mirror.

"Princess, you had your spleen removed less than two weeks ago. Are you sure you even want to go to this tonight? Nobody would blame you if you wanted to skip it. We could stay in and watch the ball drop from the couch with Myrtle." Easton moves behind me and runs his hand over my bare back. "I'm sure I could find plenty of doctor-approved, non-strenuous activities to keep us busy."

I turn to face him and adjust his black bow tie. "Easton Hayes. I am wearing a lace and silk, hand-dyed pink panty and strapless bra set that cost more than some people's mortgage payments. At the end of this night, we're going to partake in all the doctor-approved activity we can. But in the meantime, I'm going to walk the black carpet in an Everly Sinclair original with my unbelievably hot husband next to me." I look him over and lick my lips. "You really do clean up nicely, husband."

I sway in my pink watercolor ballgown that was hand-sewn by Evie. A small, pink satin ribbon ties around my neck, leaving my back completely bare. Add to that, this stunning gown cinches in at the waist and flares out in box pleats, and it's the most gorgeous gown I've ever seen, and my incredibly talented best friend made it just for me.

"We're going to show the world we're all right, and I'm going to start a campaign to change the privacy laws in this country. I don't want anyone else to ever have to go through what we just went through. And with Becket's help, I don't plan on stopping until we've got the change we want. One day, we're going to have babies, E. And I don't want them to have to deal with this too, just because they're the children of a fabulously skilled hockey player and a mother who hit the genetic lottery. We're no different from anyone else, and I want our children to be safe and able to live a normal life like everyone else."

"You are incredible, Madeline Hayes." He takes my face in his hands. "I love you, baby."

"Till we're old and gray, hockey boy. Till we're old and gray."

We ride over to the event in a limo, which Crew rides in with us. He has two other men inside the ballroom already. Not to mention the men Sam has here, simply because his wife and my siblings are here. We've all doubled down on security since the accident. It was too close for any of us to feel comfortable.

Some of the media outlets have backed off.

For now.

The Kroydon Kronicles posted a statement after the accident, saying they would no longer run any images gotten by nefarious means. I not sure exactly what constitutes *nefarious*, but they've been willing to leave us alone.

For now.

I doubt that will last though.

The Ballroom at the Beacon House is one of the oldest ballrooms in the city. It overlooks the silver glittery ball the city drops at midnight and provides a perfect view of that and the fireworks from the roof. My sister-in-law, Daphne, has been running this event for years to raise money for her charity supporting the youth of Philadelphia, so it's no surprise at all that all my friends and family will be there tonight.

We coordinated with the girls and Maddox and Callen so we all arrive together. We walk the black-velvet carpet behind them but don't answer questions unless they're asking who I'm wearing, and then I happily name drop my best friend. But one reporter looks familiar. I think I've seen her at Easton's games. "Lindy, how are you feeling?"

I link my arm through Easton's good arm and pose as flashes go off. "I'm feeling incredibly lucky, thank you for asking. I'm lucky to be alive. I'm lucky to have the love of my life next to me. Not everyone involved in our accident was as lucky as us—all in the name of a photo." I consider walking away but don't. Instead, I stand there with my strength at my side, holding me up. "I thank you all for being here tonight. For helping to bring awareness to this incredibly important cause. And I ask you to consider what photos you use in your magazines, newspapers, blog posts, and TV shows. I will always stop for you at events. But I beg you to consider what you're doing and how you're doing it the rest of the time."

Easton moves his hand to the center of my back and guides me into the event. "You feeling okay, princess? You need to sit down?"

"I'm okay for now, but I'll sit soon. I promise."

We slowly make our way over to one of the many tables my family has reserved, and eventually take our seats for dinner and Daphne's speech. Waiters in tuxes with tails and white gloves pass out champagne, but Easton and I stick to soda.

Medication and alcohol is a bitch.

Eventually, Kenzie moves next to me. "How are you feeling?"

"Other than being done with people asking me that?" I answer. "I'm okay. Tired and sore, but I want to stay until midnight. This Cinderella wants to kiss her prince."

"Just try not to lose your shoe. Your balance hasn't been the best lately."

"Brat," I tease and glance down at my phone to check the time.

"Hey, did you ever watch that video I sent you a few weeks ago? The one I told you to watch with Easton?"

I try to remember what the hell she's talking about but can't. I guess I didn't watch it. "I don't think I did."

"Do me a favor, scroll back in our messages and watch it for me before midnight." She sips from her glass of champagne and smiles.

"What about watching with Easton?" I ask.

"Don't worry about it. Just watch it now. I'll see you in a few." Then she gets up and leaves me staring at my phone.

I look around for Easton but don't see him in the crowd of tuxes and gowns.

It only takes me a minute to scroll back and find the video, and when I press play, I choke on a cry.

Kenzie has the camera in selfie mode. "I hope you guys are still watching this when we're all old and decrepit and going on seniors' cruises together. I love you guys." The camera flips and pans over to Easton, who's waiting at the end of an aisle. Elvis stands on one side of him, and Pace stands on the other side. Then the camera moves to me, walking down the aisle on Maddox's arm.

Oh my God.

She taped it.

The whole thing.

From our silly vows to the look on Easton's face when Elvis said he could kiss his groovy chick.

"She finally showed you that, huh?" Madman moves behind me and watches over my shoulder. "You know, I'd have talked you out of doing it if I didn't think it was what you really wanted to do. Or if he was a douche. We don't deal with douches, right?"

I sniff. "Right. No douches," I agree with Maddox.

"Come with me." He holds his hand out for me. "Easton's waiting for you on the roof for the ball-drop."

"That's half an hour away," I argue. "And it's cold."

"They've got so many heat lamps up there, I'm surprised the fire inspector hasn't shut this place down. Come on. We'll take the elevator. I don't want you falling on your ass on the stairs, trouble."

I know it's no use fighting with Maddox, so I place my hand in his, and we take the elevator up to the roof. When the doors open, Brandon is standing there, waiting for us. "Thanks, Maddox," Brandon tells him, and Mad turns and kisses my temple, then walks away.

"Big guy?" I question as Brandon curls my arm around his.

"Do you have any idea how incredibly proud I am of the woman you've grown into, shortcake?"

"What's going on?" I ask, and he smiles so big, I think it might split his face open.

My mother comes around the corner and hands me a bouquet of the palest pink roses I've ever seen. They match my dress perfectly. "Mom?"

"You wanted small. You wanted no fuss. We're giving you both. Your family and closest friends are around the corner. No fuss. No muss. Everyone was going to be here for the gala anyway. We just hijacked the roof for a few minutes."

"Mom . . . I don't know what to say." My voice shakes as I take the flowers and throw my arms around her.

"Don't say anything, darling. Just be happy and let me watch you marry that man. Because, Madeline, no on in the world will ever love you the way Easton Hayes does. When you were hurt . . . My God, he loves you."

She wipes my face, then takes gloss out of her purse and dabs it on my lips.

"I'll see you after, sweetheart."

Brandon watches Mom go, then walks us around the corner to where my family members all stand on either side of a black-velvet aisle. And there, at the end of the aisle, is my husband, standing next to my brother, Max, who'd gotten ordained to marry one of my siblings at some point. I laugh because I can't even remember which one.

"You ready?" Brandon asks.

"If I could, I'd run to him right now," I whisper back and lock my eyes on my husband.

Once I reach the end of the aisle, Brandon puts my hand in Easton's, and I lean in and run my lips over E's mouth. My family laughs, but I don't care. "I love you, hockey boy."

"Till we're old and gray, princess."

*T*wenty minutes later, my husband's arms are wrapped around me as we watch the ball drop and yell along with the countdown.

Ten.

Nine.

Eight.

Seven.
Six.
Five.
Four.
Three.
Two.
One.

"Happy New Year," I whisper as Easton cradles my face in his hands and presses his lips to mine.

"I told all of you they'd be kissing when the ball dropped at midnight," Everly exclaims right before Maddox wraps an arm around her back and dips her dramatically, then plants a serious kiss on her lips.

My entire friend group stares as he stands her back up and wipes his mouth.

"Happy New Year, madman," Evie kinda pants, and Maddox smirks.

"Not happening, demon spawn." He turns to Gracie, and she shoves him back.

"Not a chance, madman."

The fireworks explode overhead, and Easton nuzzles my ear. "How about we go home and have our own fireworks display?"

"Yes, please." I rest my head on his chest and wrap my arms around his waist. "Can I tell you a secret, hockey boy?"

"Anything, princess."

"You were my birthday wish."

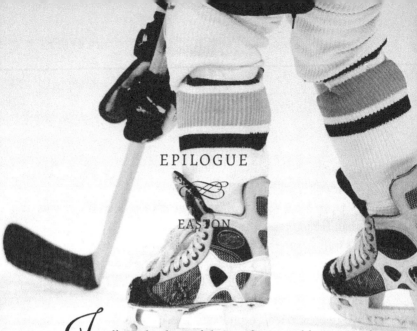

EPILOGUE

EASTON

I walk in the door of the condo, so goddamn proud of myself.

It's not like it's easy finding a gift for the woman who could buy a small island if she really wanted to. Not to mention, my wife doesn't like presents, and she's not the biggest fan of candy. But she does love cupcakes. Thank you, Amelia. I stopped by Sweet Temptations and picked up a dozen Valentine's day cupcakes, in pinks and reds with sprinkles and hearts, then grabbed a nice bottle of her favorite champagne and ordered her a few pairs of her favorite soft fuzzy socks.

Pace told me it was the shittiest Valentine's present he'd ever heard of, after he asked me if I was in high school again and then informed me jewelry or lingerie was the only way to go.

I pointed out just how single he actually is and asked him how long it's been since he had a girlfriend for Valentine's day. When he couldn't remember, I decided I won the argument and went with my plan.

My plan did not, however, include my wife crying hysterically on the couch.

"Princess . . ." I kick the door shut, cross the room, and squat down in front of her. "What's wrong?"

"You. It's all your fault." She throws her tissue on the table, where a pile of other crumpled tissues lay. "You've always got to be the best at everything. Everything," she sobs. "Just once, couldn't you be mediocre?"

"Baby, I'm gonna need you to help me out here and tell me what you're trying to say because I'm not following." I put her presents down on the table next to the gross tissues. "Want a cupcake?"

She eyes the pink Sweet Temptations box and nods.

Then she looks at the chocolate cake with pink buttercream icing, like it's betrayed her, and wails, "My ass is going to get so fat."

"It's one cupcake, baby. It's not going to make your ass fat." Holy shit. What the hell happened to my normal, sane wife, and when can I please get her back?

She rips the top off the cupcake, flips it upside down and smooshes it on the bottom half. Then I watch as she eats the entire thing in two bites. Damn.

"My boobs are going to get big too. I guess that's a positive. But I swear to God, Easton . . ." She gives up halfway through her rant and grabs another cupcake.

"Madeline, I'm not sure if I should be scared or if I should just take away the sugar. Want to help me out here?"

Lindy's stormy-blue eyes lock with mine.

"You knocked me up."

I tilt my head and stare at her as the words sink in. "What?"

"I'm pregnant. We're due in September. I saw my brother's wife, Wren, today. You know, the ob-gyn. Do you know what it's like to have your sister-in-law look in your hoo-hah and tell you there's a human up there?"

"You're pregnant?" I ask in absolute shock.

"Yup. We're due September twenty-second, to be exact."

"Holy shit," I laugh and drop to my knees, then pull her to the edge of the couch. "You're pregnant? We're gonna have a baby?"

"Yeah," she nods her head and finally smiles. "We're gonna have a baby."

"Baby . . . you're gonna have my baby." I drag her into my lap and kiss her with every unbelievable emotion flowing through me.

"If the baby's head is as big as yours, I want a divorce," she laughs through her tears. "Not really. But if their head is huge, you owe me big-time. Cupcakes for the next nine months."

"Anything you want, princess. I'll give you anything you want."

She runs her hands along my face and through my hair as her eyes dart between mine. "You've already given me everything, Easton. Everything I could ever dream of. You're really not upset?"

"I don't think I've ever loved you more." I run my hand under her shirt and over her stomach, in complete awe of her.

Lindy wiggles in my lap and tears her shirt over her head. "Oh, fuck it. It's not like you can get me more pregnant."

I stand up with her in my arms, liking the sound of that. "No, I can't."

"Promise you'll love me, no matter how big my ass gets?"

"Your ass will always be perfect, baby." I squeeze it for good measure and drop her onto the bed.

"Ohh. You're good."

"Just wait and see how good I can be, princess."

I was good four times that afternoon and three more that night.

And I did not get her more pregnant.

The End

Want more Madeline & Easton?
Download their extended epilogue here!

Download the extended epilogue here

The Philly Press

KROYDON KRONICLES

NOT READY TO SAY GOODBYE YET?

Looks like our favorite Kroydon Hills socialites are at it again. Baby Kingston got her happily ever after, now is the blonde bombshell, Everly Sinclair, looking to settle down too?

The Wildcat, Everly's book, is releasing February 15th.

Preorder The Wildcat Now

#KroydonKronicles #TheWildcat #BlondeBombshell

WHAT COMES NEXT?

If you haven't read the first book in the Kings Of Kroydon Hills series, you can start with All In today!

Read All In for FREE on KU

ACKNOWLEDGMENTS

No book has ever been harder for me to write, than Easton and Lindy's book. I couldn't wait to get my hands on these two, but life got in the way, as it has a habit of doing. Once I was finally able to dig in, I was lucky enough to be surrounded by an incredible group of women and one man, who made this happen.

As always, to my husband, without you, this book would not have been written. You are a stronger man than most people will ever get the chance to know outside of a book. I love you and thank you for everything you sacrifice to keep our family functioning while battle to meet my deadlines.

To my dream team, Bri & Heather. I love you and am so grateful for you ladies. There are no words big enough.

Tammy, Jenn, Vicki & Kelly, your honest feedback gave Easton and Lindy life. Thank you so much for all of your help.

And Jenn – Second Gen math is impossible. Thank you for making it work for us.

My editor, Dena - We did it! I was really worried about this one.

For all of my Jersey Girls ~ Thank you for giving me a

safe space and showing me so much grace. I can't tell you how much I appreciate you and your love of this world.

To all of the Indie authors out there who have helped me along the way – you are amazing! This community is so incredibly supportive, and I am lucky to be a part of it.

Thank you to all of the bloggers who took the time to read, review, and promote The Keeper.

And finally, the biggest thank you to you, the reader. I hope you enjoyed reading Easton & Madeline. If someone had told me, when I introduced her character as a toddler, 12 books ago, that I would write 16 books, and this little girl would grow into a strong woman, I would have laughed. I hope you love them as much as I loved being lost in their world.

ABOUT THE AUTHOR

Bella Matthews is a USA Today & Amazon Top 50 Bestselling author. She is married to her very own Alpha Male and raising three little ones. You can typically find her running from one sporting event to another. When she is home, she is usually hiding in her home office with the only other female in her house, her rescue dog Tinker Bell by her side. She likes to write swoon-worthy heroes and sassy, smart heroines. Sarcasm is her love language and big family dynamics are her favorite thing to add to each story.

Stay Connected

Amazon Author Page: https://amzn.to/2UWU7Xs
Facebook Page: https://www.facebook.com/bella.matthews.3511
Reader Group: https://www.facebook.com/groups/599671387345008/
Instagram: https://www.instagram.com/bella.matthews.author/
Bookbub: https://www.bookbub.com/authors/bella-matthews
Goodreads: https://www.goodreads.com/.../show/20795160.Bella_Matthews
TikTok: https://vm.tiktok.com/ZMdfNfbQD/
Newsletter: https://bit.ly/BMNLsingups
Patreon: https://www.patreon.com/BellaMatthews

ALSO BY BELLA MATTHEWS

Kings of Kroydon Hills
All In
More Than A Game
Always Earned, Never Given
Under Pressure

Restless Kings
Rise of the King
Broken King
Fallen King

The Risks We Take Duet
Worth The Risk
Worth The Fight

Defiant Kings
Caged
Shaken
Iced
Overruled
Haven

Playing To Win
The Keeper
The Wildcat (coming soon)
The Knockout (coming soon)

The Sweet Spot (coming soon)

CHECK OUT BELLA'S WEBSITE

Scan the QR code or go to http://authorbellamatthews.com to stay up to date with all things Bella Matthews

Made in United States
Orlando, FL
18 February 2025

58672384R00173